4 WEEKS

The Jade Garden

The Jade Garden

New & Notable Plants from Asia

Peter Wharton,
Brent Hine,
and Douglas Justice

in association with the
University of British Columbia Botanical Garden
& Centre for Plant Research

With the assistance of
David Tarrant, Ingrid Hoff, Daniel Mosquin,
Josephine Bridge, Judy Newton, Karen Justice,
Iain Taylor, June West, and Jane Hutcheon

Timber Press
Portland · Cambridge

Acknowledgments: We thank Susyn Andrews, Peter Cox, Martyn Dickson, Michele Funston, Chris Glenn, Howard Higson, Daniel Hinkley, Steve Hootman, Roy Lancaster, Edward Needham, and Olav Slaymaker for providing photographs and lending other valuable assistance.

Photographs: Except where noted otherwise in the photo captions, the photographs show plants growing in the David C. Lam Asian Garden at the University of British Columbia.

Published in 2005 by

Timber Press, Inc.
The Haseltine Building
133 S.W. Second Avenue, Suite 450
Portland, Oregon 97204-3527, U.S.A.

Timber Press
2 Station Road
Swavesey
Cambridge CB4 5QJ, U.K.

www.timberpress.com

Printed through Colorcraft Ltd., Hong Kong

Library of Congress Cataloging-in-Publication Data

Wharton, Peter.
    The jade garden : new and notable plants from Asia / Peter Wharton, Brent Hine, and Douglas Justice in association with the University of British Columbia Botanical Garden & Centre for Plant Research ; with the assistance of David Tarrant . . . [et al.].
        p.   cm.
    Includes bibliographical references (p.   ) and index.
    ISBN 0-88192-705-8 (hardcover)
    1. Botany, Economic—East Asia. 2. Plants, Cultivated—East Asia. 3. Plant introduction—Europe. 4. Plant introduction—North America. 5. Exotic plants—Europe. 6. Exotic plants—North America. I. Hine, Brent. II. Justice, Douglas. III. University of British Columbia. Botanical Garden and Center for Plant Research. IV. Title.

    SB108.E199W49   2005
    581.6'095—dc22

                                                                    2004020331

A catalog record for this book is also available from the British Library.

For the Honourable David C. Lam

Contents

Color photos follow pages 32, 64, 96, and 176

Preface

QUENTIN CRONK
director, UBC Botanical Garden

The University of British Columbia Botanical Garden and Centre for Plant Research is the oldest of Canada's university botanical gardens. The garden is located on the edge of the city of Vancouver, to the southwestern side of Point Grey Peninsula atop cliffs that descend to the Strait of Georgia, and is sheltered by great native grand firs (*Abies grandis*) and cedars (*Thuja plicata*). Visible from the garden over the Strait of Georgia are the mountains of Vancouver Island. Beyond that, the next landfall is the coast of Asia.

The largest continent, Asia has always been a metaphor for vastness. The poet George Barker in his touching poem to his mother wrote:

> Most near, most dear, most loved and most far,
> Under the window where I often found her
> Sitting as huge as Asia, seismic with laughter.

Huge seismic Asia, a continent of volcanoes and earthquakes, has experienced the recent uplift of the most massive mountain range and plateau on earth, caused by the northward continental drift of India and its eventual collision with the rest of Asia. The diversity of the landscape and climate of Asia has given rise to unusually rich and varied plant life. China alone has more than twice the number of native plants as the United States. It is therefore impossible for any single book to do justice to the plants of Asia.

For this book, the three UBC Botanical Garden curators—Peter Wharton, Brent Hine, and Douglas Justice—all with long experience and deep practical knowledge of the plants of temperate Asia, describe a selection of that region's plant wealth. They concentrate on those plants that are hardy in North America (at least in the Pacific Northwest) and those plants that have been studied first-hand at UBC Botanical Garden. Many familiar plants, well served by other books, are not included here. Instead, plants have been selected that are esoteric, scientifically interesting, and exceptionally beautiful.

The diversity of Asia is mirrored in the diversity of its plants. Southeast Asia is the land of the durian[1] and the rattan. South Asia is home to the feathery neem and the rhododendron forests of the cool hills of North India, Nepal, Bhutan, and Sikkim. Southwest Asia is the source of frankincense and myrrh, and (in Turkey) of the Black Sea rhododendrons that stopped Xenophon's army.[2] Northern Asia hosts the shimmering birches of Russia and Siberia. Central Asia (Afghanistan, Tajikistan, Uzbekistan, Kazakstan, and Kirghizia) has been a cradle of domestication for pistachios, pears, plums, apricots, and apples.[3] East Asia (Japan, Korea, China, and Taiwan) is the land of the camellia, peony, chrysanthemum, and maple.[4] This richness baffles as much as delights. *The Jade Garden,* as its title suggests, focuses on East Asia, reflecting the immense contribution that the forests of temperate East Asia have made to the gardens of Europe and North America.

Vancouver, throughout its history, has had strong cultural links with Asia, particularly with China, Japan, and Korea. Therefore, a major interest of UBC Botanical Garden is temperate East Asian plants. This interest was fostered mightily by Gerald Straley, research scientist and curator of collections in the Botanical Garden from 1982.[5] Gerald Straley came to Vancouver from his native Virginia, in the United States, in 1976, and obtained a PhD in taxonomic botany at the University of British Columbia in 1980. As well as being curator of collections at the Botanical Garden, he also held the position of adjunct associate in the then UBC Plant Science Department, and was research associate and director of the Herbarium in the Botany Department. Gerald had a special talent for shedding light on the complexities of taxonomy for the often mystified student. His life was tragically cut short by illness in 1997, but his legacy is still felt in the garden. *Rehderodendron macrocarpum* was his favorite garden tree, and it is fitting that a tree of the species was planted in the David C. Lam Asian Garden in April 2002 to memorialize his contributions to the UBC Botanical Garden. Straley's publications relating to the cultivation of Asian plants, including his classic *Trees of Vancouver* (1992), can be found in the bibliography at the back of this book.

Regarding the genesis of the UBC Asian plant collections, two other names must be mentioned: John Neill and Alleyne Cook. John Neill came to UBC Botanical Garden in the spring of 1971 as a research scientist, adding to his responsibilities as associate professor in the Plant Science Department. He will be long remembered by innumerable students who took his woody plant identification courses, especially those in the UBC School of Landscape Architecture, which he helped found. He was instrumental in the development of the UBC Rhododendron Species Collection, which arose from the creation of the Rhododendron Species Foundation in the United States. This collection formed the core of the present collection in the David C. Lam Asian Garden. He is also responsible for a number of notable campus plantings at UBC in the 1970s. Dr. Neill took a particular interest in the formation of the David C. Lam Asian Garden, and remained an advisor until his retirement in 1981.

Alleyne Cook has been a stalwart friend and supporter of the UBC Botanical Garden since the 1960s. He moved from his native New Zealand to England in the early 1950s to work at Sunningdale Nurseries in Sussex. He then brought his considerable horticultural talent to Vancouver, where he worked for the Vancouver Park Board for 23 years, starting in 1966. He is best known for his imaginative rhododendron and magnolia plantings in Stanley Park, Vancouver. Alleyne Cook has always been a generous source of advice and plants for the UBC Botanical Garden and his influence may be seen in our magnolia and rhododendron plantings. In 1955, he helped found the Vancouver chapter of the American Rhododendron Society.

For the genesis of this book, we must thank the Honourable David C. Lam. His generous and far-sighted endowment of the David C. Lam Asian Garden allows us to grow temperate East Asian plants in a sheltered coniferous woodland setting. The current curator of the Asian Garden, Peter Wharton, has been able to build a collection of Asian plants rivalling any in North America for botanical interest. At present, the Asian Garden is home to 3726 accessions representing 2150 taxa in 350 genera. With great pleasure, we dedicate this book to David Lam. I have asked my predecessor as director, Bruce Macdonald, to write the appreciation that follows.

We invite everyone to visit our garden either in person or online (www .ubcbotanicalgarden.org). If fortune smiles on us, we may be able to fulfill many future plans for further cultivation of, and research on, Asian plants. We hope to extend the David C. Lam Asian Garden, even to being able to grow frost-sensitive plants in an Asian House under glass, and we plan to foster exchanges of staff and students with our partner institutions in China and elsewhere in Asia. We have already benefited from our close relationships with the Hokkaido University Botanical Garden in Japan, and in China, the Nanjing Botanical Garden, the Kunming Institute of Botany, and the Kunming Botanical Garden, and we hope to build on these and similar associations in the future.

Mention should be made here of an international treaty that has relevance to recently collected wild-origin plant material that is not yet in the horticultural trade; some of the plants in this book fall into this category. Under the spirit and terms of the Convention on Biological Diversity (CBD),[6] any commercial use of wild-collected plants should involve a financial return to the country in which the material originated. The UBC Botanical Garden adheres strictly to CBD guidelines through its own CBD code of conduct (http://www.ubcbotanicalgarden.org/

conservation/cbd.php). The convention was signed at Rio de Janeiro in 1992 by most countries, including East Asian countries, Canada, and all European countries (the United States is not a signatory). The key points of the agreement are that nation states have sovereign rights over their biological natural resources, and will endeavour to make available their biological natural resources for appropriate scientific and environmentally sound commercial or social uses. In the event of commercial end use of biological natural resources, the commercial user, under the CBD, has the responsibility to ensure equitable benefit sharing with the country of origin. The idea behind the agreement was to promote the conservation of biodiversity by making nation states both responsible for biodiversity and the financial beneficiaries of it. However, the raft of poorly understood legislation that was often engendered has had in some cases the effect of suppressing the equitable and environmentally sound use of biodiversity, rather than promoting it, as was its original aim. This problem, widely acknowledged, is being tackled by a series of subsequent meetings of the convention.

Notes

1. The durian (*Durio zibethinus*) may be the only fruit dispersed by tigers. In Sumatra, the Sumatran tiger avidly eats the fruit (which smells like something between wild strawberries and blocked drains), as do other large forest animals. In parts of Indonesia, it is customary during the durian harvest to leave a certain number of fruit for the tigers.
2. Xenophon's army, when starving at Trebizond after the retreat from Babylon in 401 BC, became nauseated after consuming large quantities of *Rhododendron luteum* honey. Curiously, in 327 BC, Alexander the Great's expedition to India suffered the same fate (probably from *R. afghanicum*). Later still, *R. luteum* honey brought low Pompey's army, while campaigning against King Mithridates in 66 BC. *Rhododendron cinnabarinum*, *R. luteum*, and its allies (but apparently not other species of rhododendron) have highly toxic nectar.
3. For instance, the apple (*Malus pumila*) was apparently domesticated in central Asia, probably Uzbekistan, from wild populations (these populations were formerly distinguished as *M. sieversii*). Early apples may have been carried by humans along prehistoric trade routes between China and the Black Sea known as the Great Silk Road. See Mabberley et al. (2001).
4. There is a Chinese saying: "If you would be happy for a lifetime, grow chrysanthemums." The chrysanthemum, or *chu*, was first domesticated in China and appears in writings of ca. 1500 BC. It was taken to Japan around AD 700, where under the name *kiku* it was adopted by the emperor as his seal. And so the imperial throne of Japan came to be known as the Chrysanthemum Throne.
5. I am indebted to Peter Wharton for providing the biographical information about Gerald Straley and Alleyne Cook.
6. The Convention on Biological Diversity Web site is http://www.biodiv.org/ and the text of the convention may be found at http://www.biodiv.org/convention/articles.asp. There is a useful "Brief Introduction to the Convention on Biological Diversity" at the International Institute for Sustainable Development Web site: http://www.iisd.ca/linkages/biodiv/cbdintro.html.

Dedication to the Honourable David C. Lam

BRUCE MACDONALD
former director, UBC Botanical Garden (1987–2000)

I t is my privilege to dedicate this book on Asian plants to the Honourable David C. Lam. David Lam's love of plants began at an early age when he became aware of how the beauty of plants brings joy and serenity to people of all walks of life.

David Lam emigrated from Hong Kong to Canada in 1967, and from modest beginnings he became a respected real estate developer in Vancouver, British Columbia. Through the years, his philosophy of returning financial benefits to the community has helped many individuals and institutions.

During the mid 1970s, David Lam attended several educational courses at the University of British Columbia Botanical Garden and became very interested in the garden's development. He and his wife, Dorothy, received much pleasure from strolling through the garden, admiring the natural beauty of Asian trees, shrubs, vines, and perennials growing under the native West Coast forest canopy. He wanted to ensure that this valuable resource would be available to students, visitors, and all interested persons from around the world. And so he made a generous gift to the UBC Botanical Garden for a new education, research, and visitor's centre to be built at the entrance to the Asian Garden. Following its construction, he provided additional funding for the surrounding landscaping. In recognition of his philanthropic gift, the university named this area the David C. Lam Asian Garden. A new rhododendron bred by Jack Lofthouse in Vancouver was also given the name 'David Lam'.

David Lam has also donated to other garden and educational projects in British Columbia. One for which he will be especially remembered is his direction in transforming the gardens of Government House, Victoria, during his term as Lieutenant General of the Province of British Columbia. The development of the Rose Garden and the English Country Garden there, as well as the forming of the

Friends of the Garden group, were just some of his achievements, and remain a lasting legacy.

Today David Lam continues to apply his considerable talent in the roof garden around his apartment in Vancouver, and to pursue his interest in plants. This book is a tribute to his generosity to the UBC Botanical Garden, educational projects, and other gardens throughout British Columbia.

The East Asian Flora

QUENTIN CRONK

C hina is roughly comparable in size to Europe or to the United States, but has a much richer flora than either: three times the species richness of Europe and twice the species richness of the United States (Axelrod et al. 1996, Kubitzki and Krutzsch 1996). The richness of the Chinese flora can be ascribed partly to the region's being a cradle of recent species formation, as in the Hengduan Shan Ranges, and also to its being a museum of lineages that are extinct elsewhere—the latter implying that the region was less affected by the increasing aridity that encroached on much of the world during the late Miocene and the Pliocene periods.

The botanical distinctiveness of the East Asian flora has been recognized formally since August H. R. Grisebach proposed the botanical "Sino-Japanese Region" in 1873. The East Asian region is distinguished by a number of endemic plant families. These families include the ginkgo (Ginkgoaceae), the eucommia (Eucommiaceae), and the handkerchief tree (Davidiaceae)—all with a single species restricted to China—and more widespread families such as Aucubaceae, with the single genus *Aucuba* of seven species distributed from the eastern Himalaya to northern Myanmar (formerly Burma) and north to the Russian Far East (Wu and Wu 1996).

In China the endemic genera and families are mainly found in three centres (Ying et al. 1993): the western Sichuan–northwestern Yunnan centre includes some deciduous trees but a large number of herbaceous plants indicates its transitional nature between the Tibetan Plateau in the west and evergreen forest to the east. The remarkable *Kingdonia* (Ranunculaceae), *Dysosma* (Berberidaceae), and *Eucommia* (Eucommiaceae) are among the endemics there. The diversity of herbaceous genera in the region (which comprises the Hengduan mountain system, deeply dissected by river valleys) may be due to the diversity of topography and climate.

In contrast, the eastern Sichuan–western Hubei centre is a warm, moist region rich in forest trees, including such genera as *Davidia*, *Dipteronia*, *Metasequoia*, and *Sinowilsonia*. The endemism here appears to be based on relicts being able to persist in the equable climate. Many of the plants, such as *Metasequoia*, are more widespread in the fossil record than in the modern flora (Bartholomew et al. 1983). The same is true of the warm, wet forest of the southeastern Yunnan–western Guangxi centre to the south, which includes endemic genera such as the conifer *Glyptostrobus* (known as fossils from North America and Eurasia), as well as *Sargentodoxa* and *Gymnotheca*.

Some of the richness of China must be attributed to its being at the crossroads of ecological zones and geographic regions. It connects the rainforest in the south to boreal forest in the north, drawing plants from these zones and all that are between. Furthermore, there are routes of geographical connection with Europe, India, and the tropical forests of Southeast Asia, and with North America. Lineages from a wide pool have thus been able to jostle into the Chinese flora, and many lineages have stayed put for a long time. But because the region is tectonically active and subject to enormous erosional forces, bursts of recent species formation have occurred in the new habitats formed recently by uplift and gorge-forming rivers.

One of the most striking features of plant distribution in China is the occurrence of closely related but geographically separated plants in East Asia and eastern North America (see Boufford and Spongberg 1983, Manos and Donoghue 2001, Wen 2001, and Davis et al. 2002). The deciduous forests of eastern North America are in many respects more similar to the broad-leaf forests of East Asia than to the coniferous forest of western North America. This East Asia–eastern North America disjunction encompasses so many different plants that previous workers have been tempted to seek a single historical explanation: that these are two remnant areas of a formerly widespread vegetation of the Tertiary period. Analyses have suggested, however, that each example results from different circumstances at different times, and many genera reveal a more complex pattern than is apparent at first glance. Ecological rather than historical commonalities may therefore explain the similarity of geographical patterns.

The genera found in both far-flung areas of the globe include *Acer*, *Cercis*, *Gleditsia*, *Gymnocladus*, *Hamamelis*, *Illicium*, *Liriodendron*, *Magnolia*, *Panax*, *Podophyllum*, and *Triosteum*. It appears that many times during warm periods in the past, from the Tertiary period to modern times, forest plants have been able to migrate by seed dispersal between Eurasia and North America over northern land bridges in Beringia or the North Atlantic. These periods of spread allowed a common flora to become established, to varying degrees, throughout the Northern Hemisphere. Alternating with these periods of flux, long periods of isolation prevailed during which the oceans acted as barriers, as they do now. Separation into vicariant floras then produced the unique characteristics of the regions by selective extinction and evolution in isolation.

The study of the East Asian flora, in itself and in relation to other north-temperate floras, is still an exciting and active field of research. The David C. Lam Asian Garden at the University of British Columbia provides the opportunity for this study by bringing together Asian plants of scientific interest in a university setting.

References

Axelrod, D. I., I. Al-Shehbaz, and P. H. Raven. 1996. History of the modern flora of China. In A. Zhang and S. Wu, eds. *Floristic Characteristics and Diversity of East Asian Plants, Proceedings of the IFCD*. Beijing: China Higher Education Press; Berlin: Springer Verlag. 43–55.

Bartholomew, B., D. E. Boufford, and S. A. Spongberg. 1983. *Metasequoia glyptostroboides*: its present status in central China. *Journal of the Arnold Arboretum* 64: 105–128.

Boufford, D. E., and S. A. Spongberg. 1983. Eastern Asian–eastern North American phytogeographical relationships: A history from the time of Linnaeus to the twentieth century. *Annals of the Missouri Botanical Garden* 70: 423–439.

Davis, C. C., P. W. Fritsch, J. Li, and M. J. Donoghue. 2002. Phylogeny and biogeography of *Cercis* (Fabaceae): evidence from nuclear ribosomal ITS and chloroplast ndhF sequence data. *Systematic Botany* 27: 289–302.

Kubitzki, K., and W. Krutzsch. 1996. Origins of East and Southeast Asian plant diversity. In A. Zhang and S. Wu, eds. *Floristic Characteristics and Diversity of East Asian Plants, Proceedings of the IFCD*. Beijing: China Higher Education Press; Berlin: Springer Verlag. 56–70.

Manos, P. S., and M. J. Donoghue. 2001. Progress in Northern Hemisphere phytogeography: An introduction. *International Journal of Plant Sciences* 162: S1–S2.

Wen, J. 2001. Evolution of eastern Asian–eastern North American biogeographic disjunctions: A few additional issues. *International Journal of Plant Science* 162: S117–S122.

Wu, Z. and S. Wu. 1996. A proposal for a new floristic kingdom (realm). In A. Zhang and S. Wu, eds. *Floristic Characteristics and Diversity of East Asian Plants, Proceedings of the IFCD*. Beijing: China Higher Education Press; Berlin: Springer Verlag. 3–42.

Ying, T. S., D. E. Boufford, and Y. L. Zhang. 1993. *The Endemic Genera of Seed Plants of China*. Beijing: Science Press.

The Natural Landscapes of China and Bordering Regions— A Botanist's View

PETER WHARTON

The plants celebrated in *The Jade Garden* are part of the green mantle that enfolds the more fortunate regions of the earth. The People's Republic of China was once richly endowed with forest, whose verdant fingers traced the valleys at the edge of the Tibetan Plateau before expanding fan-like across the corrugated landscape of East Asia. The residual forests of today, although much diminished, are still a rich reservoir of plant life.

The geomorphology, or landscape form, and the chemical composition of the constituent rocks are vital elements in what determines the type of vegetation that covers any one location. Other key factors in this interplay include latitude, elevation, aspect (or exposure), rainfall, and increasing human environmental influences.

In this chapter, place names loom large, particularly in relation to the many *shans*, or mountains, mentioned. Along with the liquidation of China's forests over past millennia, the most productive lowland plains and more accessible highland areas have been converted to intensive agriculture. To our dismay this destruction continues unabated in many countries in East Asia. For the tourist, gardener, student, or botanist who appreciates plants in their wild environment, the best locations to visit are within the patchwork of mountainous terrain that courses the physical fabric of China and bordering countries. Many of the plants described in this book have their origins in these mountain refuges.

Mountains, often geographically isolated from their neighbours, can have unique plants and distinctive plant communities, and this factor contributes to the remarkable plant diversity in China, as well as in other East Asian countries, including Japan, Taiwan, Vietnam, and Myanmar (formerly Burma). A number of the "sacred mountains" of China, such as Emei Shan (Mount Omei), are well known and protected, yet there are many more that are just names on a map.

Those need to be studied and appreciated by all who value the wild world. In many cases these mountain areas require more protection, if indeed Asia's "Jade Garden" is to be nurtured for the future. There are some encouraging signs that these mountains are receiving more protection, but it will take time, resolve, and restorative, ecologically based afforestation to turn the tide of destruction.

Today the garden public is bathed in the accounts of intrepid plant exploration from the "Golden Age" of plant discovery during the early twentieth century, to stories of those who continue this work today with the same sense of wonder, joy, and dedication. Perhaps no other plant explorer has combined the talents of field botanist and geographer better than the late Frank Kingdon Ward (1885–1958), who understood the inextricable link between plants and the landscape that sustains them (Cox 1945). Following, we embark on a physical journey through one of the most geologically active regions of the world, and this activity, as we shall see, is a key factor in the floral wealth of this fabulous region.

The Four Great Steps

China is the third largest country in the world after Canada and the Russian Federation with a landmass of 9.6 million square kilometres. China is surrounded by Korea to the east, Russia to the northeast and northwest, Mongolia to the north, Kazakstan, Kyrgyzstan, Tajikistan, Afghanistan, Pakistan, Nepal, Bhutan, and India to the west and south, with Myanmar, Laos, and Vietnam along the southwestern boundary. This land is dominated by rugged country, with two-thirds covered in mountains, hills, and plateaus. These highlands mesh with a diverse array of basins, plains, and prominent river systems that give rise to four major physical regions—the "Four Great Steps."

The First Step in the west is the vast elevated terrain of the Tibetan Plateau, which covers 2.2 million square kilometres and averages over 4000 metres above sea level. This plateau is the highest and largest in the world, and is often called the "roof of the world."

The Second Step from the eastern margin of the Tibetan Plateau extends to a north-to-south physiographic line stretching from the Da Hinggan Mountains in Nei Mongol (Inner Mongolia) in the north, through the Taihang Shan (southwest Shanxi), to the Wu Shan (northeast Sichuan) and the Xuefeng Shan (eastern Hunan) in the south—the line being referred to as DTWX. The region is defined by plateau and basin country, with general elevations from 1000 to 2000 metres, and with a maximum altitude of 3000 metres.

The Third Step lies east of the DTWX line and includes the major agriculturally productive plains and basins of eastern China to the coast of the Pacific Ocean.

Geopolitical Map of Asia.

Provinces and Key Mountain Ranges of China.

These plains are interspersed by low hills that are generally below 500 metres, though isolated mountain massifs can attain up to 2000 metres in height.

The Fourth Step is the continental shelf. Here, during the Pleistocene (Ice Age), lower sea levels extended the coastline by 200 to 400 kilometres into the East China Sea.

THE MAIN SCOPE of this book covers China, and to a lesser degree bordering countries such as Korea, Japan, Vietnam, and Myanmar. The volcanic geological province of the Japanese archipelago is a significant part of the Pacific "Ring of Fire" that continues south through the Ryukyu Islands and Taiwan to the Philippines. These islands frame the eastern edge of our region of focus. All have generally mountainous spines and restricted coastal plains. The altitudinal range, varied geology, geographical isolation, and maritime climate have led to great biotic diversity, but the islands share many common elements with their larger mainland neighbour to the west.

The first three of the Four Great Steps of China and bordering regions can be further divided into a total of seven separate subregions for our purposes in this book. The First Step, the Tibetan (Qinghai–Xizang) Plateau, is the first subregion. The Second Step is comprised of the Xinjiang, the Inner Mongolia uplands, and Qin Ling Shan; the Red Basin and central uplands; and the Yunnan–Guizhou Plateau and Western Mountains. The Third Step includes the eastern highlands, the Shandong Peninsula, and the northeast coastal region; the northeastern and eastern lowlands; and finally, the southern uplands, basins, and coastal plains. I encourage the reader to use the fine *Atlas of China* by Liu (1996), particularly the topographic section, and also the popular *Nelles Maps*, to navigate the formidable array of names and locations mentioned here. The most challenging aspect of writing this account has been to describe the complex regional mountain geography of this vast area. These mountains often form provincial boundaries or overlap them, sometimes in a bewildering manner.

The Tibetan (Qinghai–Xizang) Plateau

This immense plateau covers an area the size of Alaska and Maine combined. Sven Hedin, the resolute central Asian Swedish explorer, rightly describes this region "as the most stupendous upheaval to be found on the face of the planet" (Hopkirk 1980). Comprised of high, cold desert, montane grasslands, and rich alpine vegetation, the general elevation in the north ranges from the Kunlun Shan at 5000 metres and rises toward its southern rim, the Himalayas, to peaks of between 7000 and 8000 metres. The greatest single tectonic event that shaped this vast region was the titanic uplift of the Tibetan Plateau and the Himalayas starting in the Miocene, 8 million years ago.

It is important for us to have a basic understanding of this seminal event. Landforms throughout much of western and central China have been hugely impacted by this past and continuing activity. This massive uplift is the direct result of the Indian subcontinent colliding with the continental margin of Asia. Before this collision, 150 million years ago, the Tethys formed a seaway that extended from Europe to eastern Asia; it gradually narrowed and then was obliterated by this intercontinental collision. The denser, colder oceanic plate was subducted beneath the Tibetan Plateau, while the lighter geosynclinal sediments were compressed and thrust upward into the Himalayas of today.

A ghostly trace of the Tethys Sea is preserved in the east–west depression, or suture, that the present-day Yarlung Tsangpo River follows from its origin 1550 kilometres west of the Tsangpo Gorge. The eastern termination of the Himalayas is marked by a stunning 180-degree bend in the Yarlung Tsangpo River at this gorge. Here, erosion has kept pace with the astonishing 6- to 14-centimetre-a-year uplift of the Himalayas. The Tsangpo Gorge acts like a fulcrum for a profound alteration to the alignment of the Himalayas and Hengduan Shan Ranges. The mountains that arch over the Tsangpo Gorge to the immediate north of the Yarlung Tsangpo River, the Nyainqentanglha Shan, form a link with the massive Hengduan Shan Ranges to the southeast. West of the Tsangpo Gorge, the mountains have a generally west–east orientation, while to the east of the Tsangpo Gorge, the mountains curve to a more north–south axis in western Yunnan before fanning out into Indochina.

This change in the mountain alignment was a result of compressional geological forces changing from south to north to west to east as we move eastward from the Tsangpo Gorge, and a direct result of the counterclockwise rotation of the Indian subcontinent starting in the west and moving east as it plowed into Asia, starting in the Eocene and continuing to the present. The collision has also resulted in huge movements eastward and southeastward, creating massive intracontinental strike-slip faulting that has, significantly, reactivated ancient mountain ranges throughout much of eastern China and into Vietnam.

The vertical cliffs of the Tsangpo Gorge ascend to two stupendous peaks, Namcha Barwa (7757 metres) and Gyala Peri (7238 metres), marking the eastern extremity of the Himalayan Range. These peaks overshadow the stygian chasm described so well by the plant explorer Frank Kingdon Ward.[1] The intense tectonic activity of this region makes it prone to earthquake activity; an example was the Assam Earthquake of 15 August 1950, which did immense damage. The Yarlung Tsangpo River itself has rent an immense breach into the Himalayan ramparts that stretches unbroken to this point from the Indus River of Pakistan. The gap is highly significant because it permits what is called the "Southwest Monsoon" to penetrate into the interior of the southeastern Tibetan Plateau. The combination of moisture and the high protective mountains of this area has resulted in a vegetational anomaly—the penetration of valley bottom tropical rain forests to its highest latitude in the Northern Hemisphere at nearly latitude 30° north in the Medog region.

The diversity and primeval nature of the Tsangpo Gorge forests are a wonder in themselves, yet it is here where another strange natural event can occur. Huge icefalls from the surrounding massive towering peaks can be dislodged, often by earth movements, then cascade headlong through the steaming tropical forests below, before plunging into the furious Yarlung Tsangpo River. Praiseworthy indeed is the establishment of the Yarlung Tsangpo Great Canyon National Park in 1999, covering 9168 square kilometres, and the Cibagou Nature Reserve in southeast Tibet, both in this unique area. The region is home to the Bengal or Indian tiger, now well protected in a series of reserves in northern India. Perhaps 30 to 35 tigers range through southeast Tibet into western Yunnan.

To the south, the turbulent Yarlung Tsangpo River turns south and then southwest, to become the Brahmaputra of India. It flows through a wide plain, which funnels the Southwest Monsoon into the Mishimishi Mountains to the north and then southwestward over the northeast- to southwest-trending Patkai Range, part of the Indoburmese Mountains bordering Assam, Nagaland, and Manipur. These incredibly wet mountains (with 5,000 to 10,000 millimetres annual rainfall) separate India from Myanmar eastward into the remarkably biodiverse Northern Triangle Forests that extend north to the "Three Corners" where the borders of Myanmar, Tibet (Xizang), and India meet. Here the floras of Assam/India, the Eastern Himalayas, Indo-Malaya, and China mingle.

This region of challenging geography and diverse flora is associated with the indomitable Frank Kingdon Ward's explorations of 1921, 1930, and 1952; he described it as perhaps the most uncompromising mountain terrain in the world (Cox 1945). Subsequently, Alan Rabinowitz, director of the Wildlife Conservation Society, has done inspiring work in the region; he was instrumental in the establishment in the Northern Triangle Forests of the Hkakabo Razi National Park in 1998, and the Hukawng Wildlife Sanctuary in 2001, both in Myanmar's Kachin State (Rabinowitz 2004). The Hukawng sanctuary is the size of Vermont and has a population of 80 to 100 tigers. Two more protected areas connecting these two impressive reserves, including the Hponkhan Razi National Park, have been added to a contiguous area of over 31,077 square kilometres of primary forest. Dr. Rabinowitz is a world leader in the conservation of the world's large cats. Here, the Indochinese tiger and the Asian elephant are the flagship animal species of this region's rich subtropical evergreen forests. These encouraging developments must be balanced against the stark realities of general forest degradation in Myanmar. Global Witness, a British organization, has published (2003) a formidable and disturbing account concerning the uncertain future of Myanmar's forests, which should be compulsory reading for anyone who has a passion for untouched forests. This glorious intractable wilderness continues in an arc south through Kachin State to the borders of Yunnan, along the Gaoligong Shan.

To the west, where the Brahmaputra debouches onto the Assamese lowlands, the southern slopes of the Himalayas stretch through Bhutan and Nepal, a region celebrated for its rich flora and soaring mountains.

Xinjiang, Inner Mongolia Uplands, and Qin Ling Shan

The vast arid wilderness of Xinjiang forms the western bastion of this region, north of the Tibetan Plateau. It is defined geographically on the south by the Kunlun Shan and Altun Shan and on the west by the Pamir Knot. The mountainous border regions of Kazakstan, notably the Tian Shan, merge through mountainous country with Altay Shan in the north, demarcating the Mongolian border before entering the Gobi Desert on the borders of Gansu, Xinjiang, and Mongolia. The imposing Tarim Basin, to the far west, receives the drainage of the surrounding mountains before the moisture ends up in the dried-up salt lake of Lop Nur. The Tarim, or in Uygur the "converging of the waters," is surrounded on three sides by imposing mountains that form the furnace walls of the infamous sand-choked Taklamakan Desert, which in Turkic means appropriately "the region where you go in and you will not come out."

On the fourth side, the vast stony Gobi Desert stretches eastward to the horizon. Through the shimmering heat, the surrounding mountains are home to fine alpine country with a treasure trove of herbaceous plants. At the southwest edge of the Tian Shan, the small, isolated Turfan Basin is in sharp contrast to the vast steppe grasslands of the Junggar Basin stretching northwest to the Altay Shan on the Kazak border. The Gobi Desert is a region of endless aridity, caused by the enormous rain shadow of the Tibetan Plateau that extends east to the borders of Gansu Province, where the Republic of Mongolia and Xinjiang meet. This region of yellow earth stretches across northern China (T. Liu 1985). These windblown, yellow silt (loess) deposits have deeply buried the landscape in this region since the beginning of the Pleistocene. The Tengger, Alishan, and Ordos Deserts of Inner Mongolia and bordering provinces form an austere region dominated by shifting sand dunes and numerous desiccated low mountain ranges.

An extensive forest covered Nei Mongol (Inner Mongolia) in historical times, but was burned in a vast inferno during the tenth century. Subsequent fires have further increased the aridity that is creeping eastward toward the borders of Heilongjiang Province. The Ordos Desert is encircled by the great north bend of the Huang He (Yellow River). Out of this land of great aridity rise two mountain massifs that are noted for remnant montane coniferous forest vegetation, and they have fortunately received recent protection. The Helan Shan to the west of the Huang He, near Yinchuan, and the Lang Shan and Yin Shan at the northern apex of the great northern loop of the Huang He are notable for their isolated forest fragments. The eastern Gobi Desert continues in all its severity along the border of Mongolia, gradually merging with the arid steppe south of the river Herlen Gol.

The Kunlun Shan–Altun Shan merges eastward with the Qilian Shan mountain complex within the Gansu's Hexi Corridor, ultimately connecting with the Qin Ling Shan in southern Shaanxi. This region on the northeast margin of the Tibetan Plateau contains several botanically interesting mountain areas including the Qilian Shan, Nan Shan, and Daba Shan. In this arid region, fragmented remnant *Picea asperata* forests are scattered along the northern slopes of these mountains, but trees are often completely absent on the arid southern aspects.

The extreme southeast portion of the Qin Ling Shan, in southern Shaanxi, and the connecting south- to north-trending Luliang Shan, in western Shanxi, extend north to the eastern Gobi Desert on the Mongolian frontier, thus forming the eastern margin of this enormous physiographic region. The Qin Ling Shan is a broad transitional zone that divides the realms of the Huang He and Yangtze River Basins, marking climatically the northern limit of the summer monsoon and vegetationally the northern limit of the important complex mixed mesophytic forest formations of central China. The western end merges with the high arid mountains of the Hexi Corridor, gradually becoming moister and more forested in southwest Gansu (Tianshui County) toward the Min Shan.

The Qin Ling Shan includes the sacred Taibai Shan with its outstanding nature reserve. These mountains overlook the Weihe Plain, the birthplace of Chinese civilization, then north over arid dissected loess uplands to the distant Ordos Desert. These mountains are covered by extensive rich and varied forests, well known to botanical explorers such as Frank N. Meyer (1875–1918),[2] the indefatigable USDA plant hunter. The drier northern parts of the range are mantled in deciduous broad-leaf forests with oaks predominating, while the southern parts are dominated by mixed mesophytic forest elements. Here, the famous Foping National Nature Reserve protects significant forests and important habitat for the giant panda.

Crossing the Wei He to the north, the Luliang Shan in Shanxi forms a transition area of steppe and desert characterized by basins of loess and exposed rocky uplands. These mountains of over 2000 metres were once covered in mixed deciduous broad-leaf forests, particularly to the north and east. The forests are now heavily degraded and fragmented, yet a few areas have intact conifer and mixed temperate forests such as the Pangquangou Nature Reserve in the north and Luya Shan in the south. The southeast extension of the Qin Ling Shan reaches into the Funiu Shan in western Henan. The Funiu Shan is noted for rich and diverse forests now protected in a comprehensive network of conservation areas, including the Baotianman, Laojieling, Longchiman, Laojun Shan, and Shiren Shan Reserves.

The Red Basin and Central Uplands

The Red Basin of Sichuan Province is dominated by the fertile agricultural Chengdu Plain, which has been devoid of forests for millennia. This huge depression, 500 kilometres across, resembles an inverted limestone dinner plate with the centre knocked out, leaving the encircling jagged edges as marginal limestone table-top mountains so typical of this region. To the west, the western Sichuan Plateau has been elevated to about 3000 metres, forming the eastern bastion of the Tibetan Plateau. Here successive ranges of mountains with deep, often arid river canyons are aligned in a south-to-north orientation, merging with the Tibetan Plateau to the northwest. In the south, these ranges merge directly with the Hengduan Shan axis near the "Big Bend" of the Yangtze, which encloses the lofty Yulongzue Shan (Jade Dragon Snow Mountains) near the city of Lijiang, Yunnan.

This vast rugged region is dominated by successive ranges starting in the east with the Qionglai Shan, then the extensive Daxue Shan, including the great Gongga Shan (Kanggar Shan) massif (7556 metres) and finally extending to the lofty Jinping Shan in the south. Farther to the west, the Shaluli Shan covers a huge area of high, rich subalpine country, looking west toward the enormous north-to-south fissure that confines the turbulent Yangtze. This boundless wild region was the focus of much plant exploration during the Golden Age of plant collecting in the early twentieth century. Huge areas of the region have experienced grievous deforestion, which has led to extensive erosion. Disastrous floods on the Huang He (Yellow River) and Yangtze River forced the government in September 1998 to ban logging in all catchment areas of these rivers. Massive ecologically based reforestation is urgently needed throughout this tectonically active region, where landslides regularly occur even where primary forest is present.

On the southwest edge of the Red Basin overlooking the Dadu River, the remarkable limestone mountains of Emei Shan (Mount Omei), Wu Shan, and Erlang Shan dominate. All are covered with astonishingly rich forests of breathtaking beauty. Emei Shan (3353 metres) is a *locus classicus* not only for its flora, but also as one of China's most sacred mountains. The peak also deals out a punishing cardiac challenge for visitors. The Pilgrims' Path requires the visitor to ascend over 22,000 stone steps for the opportunity to be spiritually uplifted by the mountaintop scenery. Emei Shan is also associated with three great plant explorers, E. H. "Chinese" Wilson, T. T. Yu, and Professor Wen-pei Fang, who described the botanical treasures of this mountain (Lancaster 1989).[3]

Rich flora continues along the southern edge of the Red Basin, the northern margin of the Yunnan–Guizhou Plateau. This jagged edge, as we move east, includes the Daliang Shan facing the Jinsha He. Here, the Mabian-Meigu-Dafengding Nature Reserve protects some of the most valuable high- to low-elevation forests of this region, and harbors a few pairs of the giant panda. The Wumeng Shan continues to the southeast on the Yunnan–Guizhou border before extending east to the Dalou Shan (north Guizhou) and Jinfo Shan, with a fine nature reserve. The latter two mountains have spectacular tabletop crags and are home to the Chinese silver fir, *Cathaya argyrophylla*. The whole region is rich in ancient conifers and diverse, poorly understood rhododendrons.

Farther east, particularly in northeast Guizhou, the undulating karst terrain becomes less forested and in some regions displays a relentless austerity. At the eastern edge of the Red Basin, the Yangtze River hugs the southern rim and passes close to Lichuan, the home of the "living fossil," *Metasequoia glyptostroboides*, before entering the Three Gorges region. Here the Yangtze cuts through the confining Wu Shan, at the eastern margin of the Red Basin. South of the river and continuing to the Hunan border, high, rough, partially forested country is interspersed with heavily populated valleys.

The Wu Shan continues north into the heavily forested Shennongjia Forest District, noted for its rich endemic flora. This high country continues through the Jing Shan and north to the sacred Taoist Wudang Shan, where part of the movie "Crouching Tiger, Hidden Dragon" was filmed. The Qin Ling lies to the north, separated from the Micang Shan by the Han Shui drainage. The Daba Shan connects the Micang Shan to the spectacular mountains of Hubei, so well known by the plant explorers Augustine Henry and E. H. Wilson. The country to the south of these mountains is dominated by severely deforested low limestone uplands with gorges and ravines. In these remote areas, the rare conifer *Thuja sutchuenensis* has escaped extinction by thriving on exposed cliff ledges.

The Min Shan continues to the northwest on the borders of Sichuan and Gansu. This location is the centre of the World Wildlife Fund's Min Shan Giant Panda Initiative Project. A number of wildlife refuges for the giant panda have been established in the core of these mountains, including the Wanglang and Wujiao Reserves in Sichuan and the Baishuijiang Nature Reserve in southern Gansu. To the south in northern Sichuan, a series of reserves have been set aside to protect giant panda habitat, which, in some cases, effectively conserves many of this region's remaining forests. These include the World Heritage Site at Jiuzhaigou and neighbouring Baihe Nature Reserve, the Baiyang Nature Reserve near Songpan, the Fengtongzhai Nature Reserve near Baoxing, the famous Wolong National Nature Reserve, and the Tangjiahe National Nature Reserve near Chengdu. The northwest rim is dominated by the connecting high ground of the Min Shan, over the swamps of the Songpan Caodi, before merging with the Bayan Har Shan on the edge of the Qinghai Plateau. This vast region of wilderness with superb subalpine vegetation eventually connects with the northern end of the Shaluli Shan, where our Red Basin journey began.

The geographic barrier of the Tibetan Plateau to the west and the Qin Ling Shan to the north protects the Red Basin and surrounding regions from the worst effects of the intense winter cold of Siberia. The region therefore has a winter climate that is remarkably mild but with frequent dense fogs. I have heard local people make the amusing comment that "if the dogs see the sun in winter they will bark with fright." The summers are humid, with heavy rains that bathe and nurture the region's diverse forests.

The Yunnan–Guizhou Plateau and Western Mountains

This region consists of two parts. In the east, the Yunnan–Guizhou Plateau forms a huge elevated limestone tableland that was formed as a result of the same tectonic forces that uplifted the Himalayas and Tibetan Plateau in the post-Miocene epoch. To the northwest, the Yunnan–Guizhou Plateau is linked to the Tibetan Plateau by an echelon of mountain ranges along the Hengduan Shan axis, the Western Mountains. The Western Mountains fan out to the southwest and continue to China's borders with Myanmar, Laos, and Vietnam.

The Yunnan–Guizhou Plateau is a tilted tableland that diminishes in elevation from northwest to southeast, from 2000 metres elevation in central Yunnan, then falling away from 1400 to 1000 metres across Guizhou and the region's southern borderlands of Guangxi and Vietnam. The whole region is dominated by extensive vertical karst landforms, often referred to as tower (cone) karst, interspersed with undulating uplands often covered with limestone pavement. This generally treeless rock desert is punctuated by extensive dry intermontane valleys and smaller dolinas (sunken solution depressions). Subterranean drainage, caverns, gorges, and clusters of vertiginous limestone towers dot the landscape. The western margin of this region is delineated by an immense strike-slip fault zone that runs north to south from the city of Lijiang along the Lishe Jiang/Yuan Jiang (Red River) to the Vietnamese border. The original amazingly rich broad-leaf evergreen forest of southern China intermingles with the rich mesophytic forests of the Yangtze Basin. In the south these forests in turn comingle with the tropical flora of Indochina. Sadly, years of deforestation, climaxing during the Cultural Revolution, ultimately reduced these forests to scattered fragments in remote mountainous areas.

We start our descriptive journey of this eastern subregion, the Yunnan–Guizhou Plateau, near Lijiang, the famous tourist mecca in northern Yunnan. The massive northward "Big Bend" of the Yangtze encloses the formidable Yulongzue Shan (Jade Dragon Snow Mountains) to the north of Lijiang. Nestled in their shadow is the celebrated Buddhist Yufeng Monastery, world famous for a number of ancient floriferous *Camellia reticulata* trees. Travelling south, the town of Dali on Er Hai (lake) is a favorite base for visiting the Cang Shan and sacred Buddhist Jizu Shan. These two destinations have rich floras, with the latter being particularly enticing, covered by untouched primeval woodlands. From Lijiang eastward, a series of ranges follow the Sichuan border to the impressive Gongwang Shan, where fine intact subalpine forests and rhododendron floras are prominent. Lowland areas have experienced heavy environmental damage from forest destruction and mining. The high limestone country continues on a broad front, extending east through fragmentary yet rich temperate and subalpine forests of the Wumeng Shan before crossing into Guizhou. The southern rim of the Red Basin continues east through desperately degraded hill country in western and northwestern Guizhou. The barren nature of this region is relieved only by a remarkable rhododendron forest stretching in a belt 50 kilometres wide in the Qianxi-Dafang-Jinpo area. This open region then merges with the forested ridges of Dalou Shan, once a refuge for the South China tiger before the 1950s.

Located in these mountains is the Dashahe Cathaya Reserve, near Daozhen (Yuxi), in solemn tabletop limestone mountains close to the Sichuan border. In addition, wonderfully dank forested limestone gorges are typical of this whole region, which are home to a host of rich vegetation. The reserve was set aside to protect populations of the celebrated Chinese silver fir, *Cathaya argyrophylla*. *Cathaya* by nature favors exposed limestone crags away from the competition of aggressive, moisture-loving, broad-leaf evergreens. These discrete populations continue north over the Sichuan border to the Jinfo Shan Nature Reserve, in Nanchuan County. About 100 kilometres to the southwest along the border, near Chishui, a number of other fine nature reserves have been established to protect tree ferns and extensive rich mesophytic forest that cover this border region. The region is being assessed for the possible future reintroduction of the South China tiger. East of Daozhen, near Wuchuan, grows perhaps the most authentic wild population of *Ginkgo biloba*. This population was only recently verified by Peter del Tredici (pers. comm. 2004) of the Arnold Arboretum, Jamaica Plain, Massachusetts. The area immediately to the south of the Red Basin rim is dominated by a broad, undulating karst lowland that extends from the Kunming region, northeast to Guiyang in central Guizhou. This barren, forbidding region is generally destitute of trees and punctuated only by clusters of bare, beehive-shaped limestone cones. Despite this at times numbing surface monotony, spectacular cav-

erns and passages lie hidden below ground. Also in the southwest corner of
Guizhou, east of Xingyi, including the Maling River Gorge, magnificent karst landscapes with numerous residual forest fragments are prominent.

The eastern edge of the plateau is more mountainous, with forest fragments
in the Wuling Shan on the Guizhou–Hunan border. Here the United Nations
Educational, Scientific, and Cultural Organization/Man and the Biosphere Program (UNESCO/MAB) has established the Fanjing Shan World Biosphere Reserve, a forested island of great natural beauty, famous for its dove tree (*Davidia*)
forest and golden monkeys, and once home to the giant panda. The low Foding
Shan then extends southwest from the Fanjing Shan to Yuntai Shan, near Shib-
ing, where a beautiful rich forest is set among dramatic limestone pillars and columns. Southward still, we pass through dissected upland country with intensive
agriculture along the valley bottoms and degraded fragmentary forest and scrub
on the hillsides. Just east of Kaili in southwest Guizhou lies the outstanding
Leigong Shan Nature Reserve. The terrain outside of the reserve is dominated by
near-vertical slopes that have been expertly terraced over hundreds of years by the
local Miao and Dong peoples. The rich forests of this mountain are famous for
their population of the redwood relative *Taiwania cryptomerioides*.

This upland terrain merges south into the indistinct low mountains of the
Miao Ling, which form the western wing of the Nan Ling of southern Hunan. The
depredations of Mao's "Great Leap Forward" in 1957 are everywhere to be seen
in this region, with scrub replacing what was once high forest, now only a mem-
ory in the minds of local elders. The Yunwu Shan southeast of Guiyang is just one
of many local mountains denuded of their green cover that cry out for tree planting and general ecological rehabilitation. This upland country becomes increasingly forested along the Guangxi border, notably in the Libo area and east into
the Jiuwanda Shan. This mountain range contains the superb Maolan ("beautiful orchid") Nature Reserve in Guizhou, which covers over 130 square kilometres of forested, jagged, karst mountains covered in primeval forests.

Over the border in Guangxi, the Jiuwanda Shan Reserve contains extensive
subtropical to cool-temperate forests. The southwest rim of the Yunnan–Guizhou
Plateau then cuts across northwest Guangxi, with numerous limestone and granite mountains coursing the region. We start from the Yuanbao Shan in the northeast, close to the Guizhou border, go southwest across the Duyang Shan, then
finally to the Liuzhao Shan on the Yunnan–Vietnam border. The Liuzhao Shan
has remarkable karst scenery as astoundingly whimsical as the tourist hotspot of
Guilin. Here the limestone towers, cones, and narrow valleys are covered in dense
vegetation, with an abundance of primary forest. In the border regions of Guangxi
sandwiched between Yunnan and Guizhou, low mountainous areas such as the
Cenwang Lao Shan and Jinzhong Shan are notable for rare tree genera, such as

Plate 1. An ancient *Acer palmatum*, 19 m high, Naejang San National Park, Cholla Pukto, South Korea. August 1985. Photo by Peter Wharton

Plate 2. Two young children beneath a *Celtis sinensis*, Cholla Namdo, South Korea. August 1985. Photo by Peter Wharton

Plate 3. Sohuksan Island, off the southwest coast of South Korea. The windswept scrub includes *Buxus, Callicarpa,* and *Eurya.* August 1985. Photo by Peter Wharton

Plate 4. Members of the 1985 Plant Expedition to the southwest coast and islands of South Korea. August 1985. Photo by Peter Wharton

Plate 5. An array of kimchi pots, near Mokpo, Cholla Namdo, South Korea. August 1985. Photo by Peter Wharton

Plate 6. The identification and preparation of herbarium vouchers at Taepung-ri, Sohuksan Island, South Korea. August 1985. Photo by Peter Wharton

Plate 7. Carved wood screen of *Iris ensata*. Yongmun-si Temple north of Seoul, South Korea. August 1985. Photo by Peter Wharton

Plate 8. Granite pinnacles and towers of the sacred Huang Shan, Anhui Province, China. October 1988. Photo by Peter Wharton

Plate 9. The slopes of the Xi Hai (West Sea) from Dispersing Cloud Pavillion, Huang Shan, Anhui Province, China. The patchwork of fall colors here includes yellow *Hamamelis mollis* and *Lindera obtusiloba* and red *Enkianthus chinensis* and *Viburnum sargentii*. October 1988. Photo by Peter Wharton

Plate 10. The fluted, steep, granite cliffs of the Huang Shan, Anhui Province, China. October 1988. Photo by Peter Wharton

Plate 11. Lawrence Lee (U.S. National Arboretum) and Peter Wharton at Peach Blossom Stream, Huang Shan, Anhui Province, China. October 1988. Photo by Peter Bristol

Plate 12. Morning light strikes limestone cliffs in the Dashahe Cathaya Reserve, northern Guizhou Province, China. October 1994. Photo by Peter Wharton

Plate 13. Dong village, Leigongshan Mountain Nature Reserve, southeast Guizhou Province, China. October 1994. Photo by Peter Wharton

Plate 14. Nu Jiang (Salween River) above Fugong, northwest Yunnan Province, China. October 2001. Photo by Peter Wharton

Plate 15. Members of the Dulongjiang, Gaoligongshan Expedition, Kongdan, Dulong (Taron) Valley, northwest Yunnan Province, China. October 2001. Photo by Steve Hootman

Plate 16. Drung woman with facial tattoos. Dulong (Taron) Valley, Yunnan Province, China. October 2001. Photo by Bob Zimmermann

Plate 17. Members of the Dulongjiang, Gaolingongshan Expedition, close to Myanmar (Burma) and Tibet. Northwest Yunnan Province, China. October 2001. Photo by Steve Hootman

Plate 18. Torrent by 12 Bridge Camp, Qiqi Pass, northwest Yunnan Province, China. October 2001. Photo by Peter Wharton

Above right:
Plate 19. Cool-temperate montane forest at 2500 m. Qiqi Pass, northwest Yunnan Province, China. October 2001. Photo by Peter Wharton

Right:
Plate 20. Canes of *Phyllostachys heterocycla* var. *pubescens* (moso bamboo). Wuyi Shan World Biosphere Reserve, Fujian Province, China. October 2002. Photo by Peter Wharton

Plate 21. Jiu Qu River. Tian Yu Feng, Wuyi Shan Scenic Area, Wuyi Shan City, Fujian Province, China. October 2002. Photo by Peter Wharton

Plate 22. A massive *Pterocarya stenoptera* festooned with epiphytic ferns. Wuyi Shan City, Fujian Province, China. October 2002. Photo by Peter Wharton

Plate 23. Village near Xichou in the cone karst region of southeastern Yunnan Province, China. April 2003. Photo by Peter Wharton

Amesiodendron and *Handeliodendron* of the Sapindaceae (Ying et al. 1993). This complex undulating karst country increases in elevation from east to west, along the southeast-to-northwest axis of the You Jiang drainage in northwest Guangxi. Forests in general, throughout this southern region near Funing, Wenshan, to Jianshui are badly degraded by intensive agriculture and wholesale deforestation.

As we approach the Vietnamese border, forest cover tends to increase, notably in Maguan County, Exterior Gulinqing, where extensive magnolia-rich forests are present. The wonderfully diverse broad-leaf evergreen forests that remain in this region gradually merge with tropical rain forest at low elevations, notably in the Red River valley. In general, west of Wenshan, the increasing influence of the Southwest Monsoon becomes obvious, enhancing the forest cover on windward slopes throughout this region and much of southwest Yunnan. The Daiwei Shan, near Pingbian on the east side of the Red River, is another fine forest reserve with diverse flora in a good state of preservation. The terrain northwest toward the Kunming Basin becomes increasingly subdued and treeless. Dominated by red soils, gully erosion, rock desert ridges, and jagged, honeycombed hills, the landscape conspires to create a sinister spectacle for the observer especially on overcast days. The famous Stone Forest of Lunan forms an island of natural vegetation surrounded by intensive agriculture or endless rocky hill scrub complete with sadly emaciated eucalyptus plantations.

The Western Mountains subregion is really an extension of the Tibetan Plateau. This portion of the limestone plateau west of the Lijian-Lishe/Yuan Jiang (Red River) line is much deformed and broken up into a series of spectacular north- to south-trending ranges that fan out to China's southern frontier. The Western Mountains in northwest Yunnan are dominated by the Hengduan Shan mountain complex, which consists of many component ranges. *Hengduan* in Chinese means "cut across," referring to the difficulty of travelling east to west (Lancaster 1989). The rugged country of northwest Yunnan is incised by three turbulent rivers, the Nu Jiang (Salween), Lancang Jiang (Mekong), and Jinsha Jiang (Upper Yangtze). Each river passes between parallel massive mountain ranges, yet all are confined within a narrow 150-kilometre strip of terrain. The flora of this region is world famous for its astounding plant diversity, from subalpine, through warm temperate rain forests, to the unique arid sclerophyllous valley bottom flora. The combination of massive mountain uplift, a monsoon climate, and protection from continental glaciation in the Pleistocene has produced amazing diversity.

In the extreme northwest, near Deqin, two magnificent reserves have been established. The Mount Baimang Reserve to the east of Deqin is the largest in Yunnan, while to the west the famous Mount Meilixue Reserve in the massive Lu Shan Range is well known to tourists. These areas preserve superb subalpine flora and diverse forests. The Gaoligong Shan defines the physiographic, floristic, and national border of the Western Mountains with Myanmar. Commencing on the Tibetan (Xizang) border and south through the famous Dulong Jiang (Taron River) drainage to the *Taiwania* forests of the Qiqi, we enter a botanical wonderland that connects directly with the Northern Triangle Forests of Myanmar. The mountains here and to the east, including the Nu Shan, are famous for their rhododendron floras. This is the vertiginous land where plant explorers George Forrest, Heinrich Handel-Mazzetti, Francis Kingdon Ward, and Joseph Rock ranged in the early twentieth century, often in awe of the spectacular natural beauty surrounding them.

In the west the heavily forested Gaoligong Shan continues to hug the border, with Myanmar's Kashin State to the immediate west. The creation of the UNESCO/MAB Gaoligong Shan World Biosphere Reserve on the Chinese side of the border is an encouraging step, but in Kachin State, a calamity is unfolding. There, a massive program of industrial logging is under way, liquidating one of the world's greatest forest ecosystems (Global Witness 2003). South of latitude 26° north, the main Hengduan Shan mountain system radiates and divides into a region of complex geology, active uplift, and considerable seismic activity. Hot springs near Tengchong are quickly becoming popular with tourists. A number of cinder cones are also of considerable interest, notably the 600-metre Mount Daying, a volcano that was active just 360 years ago. The main mountain ranges of the south from east to west include the Ailao Shan, which forms the eastern boundary of this Western Mountains subregion. The Ailao Shan Nature Reserve is one of several protected areas in this fine linear highland that continues south into Honghe Country of the Red River region before crossing the Vietnamese border.

In Vietnam this high ground merges into the celebrated Fan Si Pan mountains of Lao Cai Province, a range known for its rich flora, especially relictual conifers. The establishment of the Hoang Long Sen National Park in these mountains will do much to protect the existing subtropical montane cloud-forests. The discovery of *Xanthocyparis vietnamensis* and an isolated population of *Taiwania cryptomerioides* has turned heads and underscores the extraordinary floral diversity of the region.

Farther west in Yunnan, the Wuliang Shan and Laobie Shan have tracts of intact forest with particularly well developed subalpine vegetation. Fragments of the original evergreen broad-leaf and lowland tropical monsoon forest occur along the Myanmar border regions. Unfortunately, deforestation in the neighbouring Shan states is extensive and well advanced.

The rich flora and benign climate of southern Yunnan has resulted in a huge variety of cultivated plants being grown there. A short list of the delights that can be sampled by a visitor to this area include tangerine, pomelo, orange, longan, lychee, and coffee. Areas to the south of Simao and Jinghong, including the famous UNESCO/MAB Xishuangbanna World Biosphere Reserve, have a genuine tropical monsoon rain forest climate. This reserve is an aggregate of several protected areas that now cover 240,000 hectares, complete with Asian elephants and the Indochinese tiger. Perhaps 30 to 40 tigers inhabit the remote mountainous country of southern Yunnan and southwest Guangxi along the borders of Vietnam, Laos, and Myanmar. In much of Indochina, these noble creatures are threatened by industrial logging, gold mining, hunting, and trade in body parts. The southern dissected mountain country continues east from Simao to the Laotian border with some fine intact forests along the Vietnam border toward Jinping and Luchan. The diversity of these mountain forests at latitude 23° north on the Vietnam border is astonishing, with small changes in altitude having dramatic impact on species composition and diversity.

The Eastern Highlands, Shandong Peninsula, and Northeast Coastal Region

The undulating high ground of this region stretches from the north banks of the Huang He through the Taihang Shan and Yan Shan to the southern foothills of the Da Hinggan Ling, forming an upland crescent around Beijing. The Qian Shan, along the North Korean border, forms the rocky spine of the Liaodong Peninsula. This highland axis then continues beneath the Bo Hai Sea to reemerge as the Shandong Peninsula, which rises to the sacred Tai Shan. Much of this region overlooks the fertile eastern lowlands and is known for its poor soils and poverty. The original extensive deciduous broad-leaf forests were stripped millennia ago, and despite recent afforestation, the region remains a land of scrubby dry-land forest at best.

The Taihang Shan forms an eastern perimeter overlooking the northern plain of the Huang He. Individual mountain massifs are host to fragments of the original transitional mixed northern hardwood and mixed mesophytic forests, particularly in the southern Shanxi, such as Zhongtiao Shan Nature Reserve (Li Shan), Jiyuan Nature Reserve, and Taiyue Shan. Farther north, the sacred Wutai Shan was once covered in superb forests, but these were gone by the late nineteenth century, except around Buddhist and Taoist temples. The same denuded rock desert with scrubby vegetation continues northwest of Beijing toward the Great Wall and Inner Mongolia. This stark barren country, home to seasonal dust storms, continues east through the Yan Shan and is relieved only by two heavily forested reserves in the adjacent Wuling Shan. This high country descends eastward into the lower Liaohe Plain thence to the heavily deforested Qian Shan and Liaodong Peninsula.

The Shandong Peninsula is well known for mountainous terrain, poor soils, and a highly indented coastline. Fine remnants of the original oak-dominated mixed mesophytic forests are found on the sacred Tai Shan and surrounding mountains including Meng Shan and Lu Shan.

The Northeastern and Eastern Lowlands

Two major plains dominate this region: the Manchurian (Dongbei) Plain to the northeast of Beijing, drained by the Liao He and Songhua Jiang, and the North China (Huabei) Plain to the south of Beijing, crossed by the Huang He and the Yangtze.

The Manchurian Plain is the most northerly plain and is now heavily cultivated and has lost most of its forest cover, particularly in the south. This plain is embraced on the northwest by the Da Hinggan Ling Mountains of Inner Mongolia. These mountains stretch to the Heilong Jiang (Black Dragon or Amur River) on the Russian border. This enormous region is home to extensive montane coniferous forests dominated by spruce (*Picea*) and fir (*Abies*), grading into mixed northern hardwoods in the south and lowland areas. Birch (*Betula*) predominates in the north, while maple (*Acer*) and linden (*Tilia*) are more abundant in the south.

In a region plagued by increasing aridity, fires are becoming a grave concern. In 1987 a cataclysmic fire broke out that extended from the northern end of the Da Hinggan Ling Mountains, hopping over the Heilong Jiang into Russia. The rampage consumed an area larger than Scotland, before fall rains extinguished it. This event is just one in a long history of forest destruction in this dauntingly vast region. Harrison Salisbury's harrowing account of *The Black Dragon Fire* (1989) is grim reading indeed and, perhaps, a vision of future calamities.

Much of northeast China (Manchuria) and the Heilong Jiang region is home to the Siberian or Amur tiger. Tigers filter southward into China and often fall afoul of hunters and of habitat destruction. The establishment of the Felidae Breeding Centre, Hengdaohez, and a huge tiger reserve in the Wanda Shan, close to the Russian border, are starting to reverse the steep decline in Siberian tiger numbers. The establishment of the Hunchun Tiger–Leopard Reserve in Jilin Province close to the Russian border is another positive step. The protection of big cats, pandas, or other large animals and the preservation of large tracts of forest go hand in hand.[4] Coniferous forests continue southeast along the Xiao Hinggan Ling south of the Amur River before crossing the Songhua Jiang into the northern end of the huge Chanbai Shandi uplands. In the 1920s, this high country was covered in old-growth *Picea-Abies* and *Pinus koraiensis* forest containing massive trees up to 50 metres high. There are remnants today of these forests in the Mudan, Woken, and Tangwang drainages of the Songhua Jiang. This river then flows through a region of larch swamps before joining the Heilong Jiang. These superb mixed hardwood-coniferous forests, with some trees of massive size, continue into the magnificent Sikhote Alin mountains on Russia's Pacific coast. A number of UNESCO/MAB world biosphere reserves have been set aside to protect these remarkable forests, notably the Bikin Valley, yet extensive industrial logging is under way in several regions. These forests then become increasingly fragmented south of the Wanda Shan, through a series of low mountains to the north and along the Russian frontier. Old-growth montane coniferous forest and rich mixed northern hardwood forests are confined to reserves such as in volcanic mountains of the UNESCO/MAB Changbai Shan World Biosphere Reserve and Lao Ling on the North Korean border. Sadly, the last Siberian tiger in this region was killed in 1980.

These forests then continue in a degraded state through the Hamgyong Sanjulgi and Puksubaek-san mountain regions of North Korea and south along the Taebaek Sanjulgi into South Korea. Forest cover merges southward into fragmented transitional, rich mixed mesophytic forests, with even some evergreen broad-leaf elements in the extreme south (Cheju-do). Japan lies to the east as part of a volcanic island chain continuing through the Ryukyu Islands to Taiwan, all with distinct, rich, ligneous (woody), subalpine floras.

The North China Plain is drained by two massive rivers. The Huang He (Yellow River) flows north of the Qin Ling Shan from the plateau hinterland of Qinghai. To the immediate south, it is connected by the manmade Grand Canal to the Chang Jiang (Yangtze River). This river arises on the Tibetan Plateau before cutting a way through the Hengduan Shan to the Red Basin, where it is impounded by the Three Gorges Dam before flowing into the Yellow Sea. The rich lowland forests of this region were decimated long ago. An imaginary return to this region 8000 years ago would reveal forests of great richness and an abundance of wildlife.

The Yangtze River marks the transition between the rich mesophytic forests of southcentral China and the deciduous broad-leaf forests of the north. These forests are typical of parts of northern Hubei, the southern Qin Ling Shan, and northern Red Basin, especially the Min Shan. Upland areas surrounding this plain are covered in forest remnants, which gives us an inkling of the original forest cover of this region. The Dabie Shan on the borders of Anhui, Henan, and Hubei is a good example, with a number of significant protected forests including the Mazong Ling, Jigong Shan, and Dongzhai Nature Reserves. The Yuntai Shan of northern Jiangsu, the sacred Huang Shan in southern Anhui, and Lu Shan, Mao Zedong's famous summer mountain retreat in northern Jiangxi, have significant intact forests that are worth visiting for their flora and fauna.

The Southern Uplands, Basins, and Coastal Plains

This topographically varied region is dominated by the east-to-west mountainous axis of the Nan Ling. A narrow physiographic, climatic, and floristically rich transition zone, it consists of numerous southeast- to northwest-trending ranges often in echelon. These start in the west, with the Yuecheng Ling on the Hunan–Guangxi border, then move east to the Dayu Ling on the Jiangxi–Guangdong border. This hill country continues into heavily cultivated lowlands along the Guangdong borderlands, eventually connecting the Nan Ling to the main south-to-north Wuyi Shan axis along the Jiangxi–Fujian border. The Nan Ling with the Wuyi Shan axis forms the spine of these two distinct regions. To the north are the mountains and basins of Hunan and Jiangxi Provinces. To the south and east of the Nan Ling, the coastal plains and uplands of Zhejiang in the far north stretch south through Fujian, Guangdong, and Guangxi.

The Nan Ling at latitude 25° north forms the southern base of a horseshoe-shaped upland spine containing the Dongting Basin Plain in Hunan and the Poyang Basin in Jiangxi, each with rivers that flow north to the Yangtze. Journeying from east to west, to the western end of the Nan Ling we pass through a series of mountain ridges, covered in rich forests. These include the Haiyang Shan, north of Guilin to the Yuecheng Ling, Mao'er Shan, and Yuanbao Shan, close to the Guizhou border. Here granitic intrusions often form isolated acidic massifs in a sea of limestone. These dramatic, beautiful mountains are covered in widely diverse forests, home to perhaps 30 rhododendron species, which is but a sample of this region's floral wealth. The Longsheng Hot Springs National Forest Park is a particularly interesting area to visit.

North into Wugang County, Hunan, the high ground connects with the Xuefeng Shan. The western edge of this upland region hugs the border with Guizhou, the margin of the Yunnan–Guizhou Plateau. This undulating karst country merges northward with the Wuling Shan, a broad northeast-trending mountain range, famous at one time for its tiger population. Here forest destruction has been severe, especially on the Guizhou side of the border. Forest cover increases on the Hunan side, becoming impressive in the Wulingyuan region of northwest Hunan. This popular tourist destination is known for its surreal karst mountains often forming spectacular groups of horizontally stratified limestone towers. The rich mesophytic forest hugs improbable crags and near-vertical rock faces, before descending into narrow ravines and box canyons. The region is critical habitat for the South China tiger. An extensive new conservation zone consisting of the Huping Shan, Taoyuan, and Badagong Shan Reserves and the Zhangjiajie area has been set up to protect the few remaining tigers in northern Hunan. The popular Zhangjiajie area and Badagong Shan Reserve have superb forests rich in wildlife. Travelling east across the Dongting Plain south of the Yangtze River, we arrive at the Three Corners region of Hunan, Hubei, and Jiangxi, where a spine of northeast-trending mountains demarcates the Hunan–Jiangxi border.

The Mufu Shan crosses the Three Corners region, where the famous sacred Taoist Mount Jiugong is located among fine forest scenery. To the south, patches of the original mesophytic forest cover the Jiuling Shan and Luoxiao Shan. The mountain spine of the Luoxiao Shan follows the Hunan–Jiangxi border and con tains a major South China tiger conservation zone connecting five nature reserves southward to the Chebaling Nature Reserve in Guangdong. This includes Yihang County in Jiangxi, where some farmland is currently being converted back to high forest to enhance tiger habitat. This country is undulating, with fragments of the original evergreen broad-leaf forests interspersed with disturbed forests of Masson's pine (*Pinus massoniana*) below 800 metres. The Jinggangshan in the Wanyang Shan is a major tourist destination for Chinese and, recently, for Westerners. Spectacular, accessible mountain scenery includes magnificent forests and historically significant sites. It is here where Mao Zedong's Long March commenced in 1934, at a low ebb in his fortunes. These mountains then connect southward directly with the Nan Ling on the Guangdong border.

The Nan Ling extends west in a complex series of individual mountain groups including the Zhuguang Shan, Yao Shan, and Jiuyi Shan on the Hunan–Guangdong border to the Dupang Ling on the Hunan–Guangxi border. This remarkable region is one of the most significant areas in China for biodiversity. It is a popular destination for professional ornithologists and amateur birders, as the forests are extensive, diverse, and often well preserved. A number of world-class reserves are located in this provincial borderland region, notably the Mangshan and Qianjiadong Reserves in Hunan, the celebrated Huaping Cathaya Reserve in Guangxi, Jiulian Shan Reserve in Jiangxi, and Babao Shan in Guangdong. The Nan Ling is prime habitat for tigers, since old-growth forests and prey are abundant. Strenuous efforts by the Chinese Ministry of Forests, ably assisted by the World Wildlife Fund for Nature and the Wildlife Conservation Society, have established tiger conservation zones connecting nature reserves that span the Hunan, Guangdong, and Jiangxi borders. The Nan Ling axis extends and then fades at the northern end of the Jiulian Shan on the Guangdong–Jiangxi border, before rising toward the Daimao Shan and Boping Ling in southwestern Fujian.

Surrounded by these mountains, the city of Longyan is home to the South China Tiger Breeding and Wild Naturalizing Research Centre, close to the richly forested Meihuashan Nature Reserve. This program envisions the eventual release of these big cats back into the extensive forested mountain wilderness area of southern China, once natural food sources are established and natural ranges are protected and expanded. In the early nineteenth century, perhaps over 4000 tigers hunted across extensive areas south of the Yangtze River. Today the South China tiger has a remnant wild population of perhaps 30 to 50 individuals in Fujian, Guangdong, Hunan, and Jiangxi. Fujian has an extensive network of protected areas within the province with over 50 reserves in excess of 500 hectares. A massive program of reforestation and enlightened forest conservation policy initiated in the late 1950s makes it a provincial leader in China and shows what can be accomplished in a relatively short time.

The Wuyi Shan follows north along the provincial boundary, defining the eastern edge of the Jiangxi-Jitai Basin, which falls away through the low Yu Shan to the Poyang Hu (lake) and Yangtze River. The Wuyi Shan becomes increasingly forested to the north, with border peaks like the Baishi Feng and Emei Feng being especially notable. The heavily forested Shan Ling spans the border with Jiangxi and merges with one of the biotic gems of China, the UNESCO/MAB Wuyi Shan World Biosphere Reserve. Viewed from the Huanggang Shan (2175 metres), the highest mountain in southeast China, ridge upon forested ridge stretches to the horizon in this reserve, home to one of the most diverse floras in China. During a visit to this reserve in the spring of 2000, there was talk of tigers being still present in the region. It is without doubt prime habitat for tigers, with over 56,000 hectares protected in a region rich in forest cover. To the east, a number of smaller reserves are notable, especially the Mandan Shan Forest Reserve, close to Nanping. Here a pristine low-elevation broad-leaf evergreen to subtropical forest occurs in a box canyon with a Buddhist temple as a destination.

The Wuyi Shan then continues north into the Zhejiang mountain region, an area of complex topography and rich tree flora. In the north, just west of Hangzhou, in the Tianmu (Jianmu) Shan, *Ginkgo biloba* appears to be growing naturally in a rich mesophytic forest. Farther south in the coastal Tiantai Shan, the forest becomes more broad-leaf evergreen in nature. Several fine reserves are located in this province including the Longtan Shan and Qinliangfeng Reserves. To the east the famous complex of four protected areas in Longquan County, the Baishanzu–Fengyang Shan Reserve is noted for its diverse forests. A few South China tigers live in this area, where several reserves are being joined together to create wildlife corridors and improve protection for these majestic creatures. This region is part of the Donggong Shan (east of the Wuyi Shan) that continues south into Fujian and forms an elevated spine that is separated from the Daiyun Shan by the Min Jiang. This river drains much of the northern Wuyi Shan and empties into the East China Sea at Fuzhou. The interior of Fujian is a subtle, romantic, undulating landscape of forested ridges and lush, intensively cultivated valleys. The interior valleys merge with the narrow submerged coastline (ria), dominated by rocky headlands, islands, inlets, and estuaries. To the east the island of Taiwan lies beyond the Formosa Strait. The distinct and varied endemic flora of this mountainous island is remarkable and is closely related to the flora of Fujian, notably the Wuyi Shan.

The coastal plain continues west into Guangdong with a flora becoming increasingly subtropical. Inland, intervening uplands such as the Lianhua Shan link northward with the Jiulian Shan to the Nan Ling. This high ground forms a crest around the northern margins of Guangdong, with the Yunkai Dashan forming the western edge of the Xi Jiang Basin, which surrounds the burgeoning metropolis of Guangzhou and the Pearl River Delta. This region of intensive agriculture and massive industrialization is close to waterways and the sea.

The upper Xi Jiang becomes the Xun Jiang, which drains the central plain of Guangxi and is bounded in the northeast by the Nan Ling and numerous upland outliers that mark the margins of the Yunnan–Guizhou Plateau north and east of Nanning. The Dayao Shan National Forest Park is remarkable for its narrow limestone columns and towers nested in subtropical to warm-temperate forests. The flora is intriguing and exemplified by a small southern outlying population of the Chinese silver fir, *Cathaya argyrophylla*, that clings to this mountain. The Shengtangshan is a provincial government resort located within the Dayao Shan, famous for impressive rock towers and rhododendron flora. This extensive and diverse forest cover extends to the Daming Shan, northeast of Nanning. Superb examples of limestone cones, called *fengcong*, or peak clusters, are a major feature to the northwest into the Duyang Shan. Rich forest fragments and associated herbaceous plants often cling to the precipitous ledges and fissures of these near-vertical features.

From Nanning the intensely cultivated river valley of the Yong Jiang (the headwaters of the Xi Jiang) gives way to the limestone uplands of the Shiwanda Shan, on the Vietnamese border, dominated again by dramatic karst landscapes. In this area the Nonggang Nature Reserve near Longzhou contains magnificent tropical rain forests, where the Indo-Malayan and Sino-Vietnamese floras mingle. Fragments of tropical forest continue northwest along the Vietnamese border into terrain increasingly dominated by superb tower and cone karst. It is especially spectacular north of Pingxiang on the Chinese side of the border to Cao Bang on the Vietnamese side. This whole border region, from the limestone tower islands of Vietnam's world-famous Ha Long Bay, through Cat Ba National Park, to the Liuzhao Shan on the Yunnanese border, has a stunning flora. Many of these individual, improbably shaped limestone mountains have rich subtropical montane forests. This terrain has ensured the preservation of many unique plants including endemics. A few Indochinese tigers make their home in this borderland region.

The Leizhou Bandao in western Guangdong is a peninsula that juts southward toward the province and island of Hainan. This island has a genuine tropical climate and once had extensive tropical rain forests. Many of these forests have been liquidated, but a number of fine reserves have been established, particularly in the mountainous southern half of the island. The Wuzhi Shan (Five Fingers) is the highest mountain massif on the island, with four extensive reserves of tropical rain forest and subtropical montane cloud forest, rich in Indo-Malayan floristic elements. Other significant reserves include the impressive Jianfengling Nature Reserve with the largest, best preserved tropical rain forest in China, with over 20,000 hectares set aside for the visitor to relish.

The Tibetan Plateau Uplift—Its Impact on the Climate and Forests of China

The uplift of the Tibetan Plateau has had a profound effect on forest vegetation in China. The vertical movements have driven the climatic development of Eurasia ever since, including the monsoon weather of eastern Asia we experience today. There is also increasing evidence that these massive geologic events since the late Miocene may have had global effects on climate, notably as a contributing factor in the onset of the Pleistocene.

The rising of the massive topographic barrier of the Tibetan Plateau has created a huge obstacle for the westerly circulation in the Northern Hemisphere, causing the jet stream to divide into two branches, one flowing around the north of the plateau and the other deflected south. The overall effect was to create a creeping and increasing rain shadow across central Asia, giving rise to the Gobi and Mongolian Deserts. The first loess deposits in northern China started just after the onset of the Pleistocene. This event appears to coincide with a time when the full effect of the Tibetan Plateau uplift started to severely obstruct direct westerly circulation. This east-to-west obstruction also effectively prevents the flow of warm, moist air from the Indian Ocean from penetrating into the interior of central Asia. The monsoon climate of Asia appears to have been initiated in the late Miocene by the rapid rise of the Himalayas. This led to progressively stronger winter and summer monsoons in Southeast Asia from the late Miocene to the Pleistocene, which in turn, led to progressive drying in central Asia.

These huge events are linchpins to an understanding of the palaeoclimatic developments that have helped shape the flora of China. China, since the onset of the Pleistocene, has never experienced the continental glaciation that occurred in Europe and North America. Nevertheless, the pulsations of glacial maxima and warmer interglacial periods have profoundly molded the natural forest cover and grasslands of China, particularly east of the Tibetan Plateau.

The impressive forests of 8000 years ago that covered the lowlands of central and southern China were a far cry from the denuded agrarian landscape of modern times. Yet these continuing landscape transformations of our recent past history are reflected in some sense by the huge, but often progressive natural changes that have ebbed and flowed across the vegetation zones of East Asia since the beginning of the Pleistocene. *After the Ice* by Stephen Mithen (2002) explores global human history from 20,000 to 5,000 BC. His descriptions of the expansion and retreat of forest types across Eurasia since the Late Glacial Maximum and a hypothetical journey down the Yangtze River are particularly illuminating.

The vegetation changes that have affected China since the Late Glacial Maximum 18,000–14,000 years ago have played a part in defining the composition of what we observe today within intact primary forests. The presence of numerous moist mountain refugia and linear north-to-south mountain ranges in China has allowed the continued survival of many ancient plant lineages from as far back as the Eocene, during a postglacial period punctuated by cold and often very dry periods. The ability of cool- to warm-temperate to tropical forest communities to move freely north and south, into and out of the main Indochinese refugia, in Vietnam, Laos, Thailand, and Myanmar in response to these changes, is of great significance. Cooling and warming events promoted the migration of whole plant assemblages along these sustaining corridors or in and out of geographically isolated mountain refugia.

This floristic interface during the Pleistocene between temperate and tropical elements underwent constant back and forth movement—the temperate sometimes having to adapt to tropical conditions, while the tropical occasionally was forced to adapt to a much cooler climate. Comprehending this factor is crucial for understanding the amazing palette of plants we see in the wilds of China and surrounding regions today. During the Pleistocene, though diminishing in the early postglacial period, the lowering of sea level created a huge lowland area in Southeast Asia (the Sunda Shelf) for forest expansion. Floristic exchange could then occur between Indonesia, the Philippines, Taiwan, Japan, and the Indochinese mainland. Vigorous mountain building continued throughout this period in western China, where the unique combination of climate, complex mountainous topography, and soils has encouraged floristic diversification in *Rhododendron* and many herbaceous plant families. The proximity of the Tibetan Plateau and a number of east–west mountain ranges in northern China, notably the Qin Ling, formed a periodic though often effective geographic barrier mitigating the worst effects of severe cold periods from penetrating into central and southern China.

During the Late Glacial Maximum, the forests of warm-temperate and subtropical latitudes of southern China and Taiwan were in full retreat, with cool-temperate forests and open woodlands predominating across the region. The Leizhou Peninsula, for example, just north of Hainan Island, in Guangxi Province, today is covered in subtropical forest, whereas it was once covered in warm-temperate pine-oak forest (Liew et al. 1998). Remarkably, most of southern China (at latitude 20° to 23° north) was covered with steppe vegetation or open grass-dominated parkland (Winkler and Wang 1993), with extensive tree communities only found on local mountains. The influence of local orographic rainfall enabled the development of these forest refugia. A cool, dry climate predominated with only weak summer monsoon activity. The climate became so dry during the period 17,000–15,000 years ago that temperate forests in southwest Yunnan almost disappeared and were replaced with dry scrub (K.-B. Liu 1991). A general retreat of rain forests also occurred, leading to replacement by grasslands and seasonal monsoon woodlands.

By 12,000 years ago, a major warming trend enabled forests to return to the north and into lowland areas as summer monsoon activity increased. Despite this, forest zones were still compressed to the south. Cool-temperate forest existed as a belt in southern China, much farther south than the present day. Central and northern China were colder than today (K.-B. Liu 1986). Extensive coniferous forests covered most of northern China, while in the south, tropical rain forests started to expand and move north.

A major climate reversal took place 11,000–10,200 years ago (the Younger Dryas period), with forests retreating in the north to the point where the Beijing area had become semidesert (Liew et al. 1995). Central China became transformed to open woodland steppe, with only patches of temperate and boreal trees surviving on moister hilltops (Zhou et al. 1992; An et al. 1991). In southwest China, generally cold dry conditions prevailed, so the forests hung on only in the wetter mountains (Jarvis 1993).

In the early Holocene, 10,000 years ago, a sudden warming and moistening of the climate resulted in extensive tree cover returning to northern Eurasia. By 8000 to 7000 years ago, conditions became much warmer and moister than the present day, with vegetation cover increasing considerably in desert areas (Winkler and Wang 1993). A general northward expansion of the temperate forest belt in China and Japan began, with much greater forest cover spreading throughout southern Asia (Cullen 1981; Zonneveld et al. 1997) during this period. This rapid climatic change is likely the result of a much more vigorous summer monsoon, compared even to that of today.

These natural events preceded those of modern humans, our civilizations, and inevitable progressive forest destruction. Human ability to destroy forests in this region is certainly an important factor in the increasing aridity of many parts of China, whether Nei Mongol or Yunnan. The forests of China act as a moisture conveyor belt for the Southwest Monsoon as it gradually moves north throughout the summer. This "precipitation recycling" (Lettau et al. 1979) is well known from research in Amazonia, but also applies to many other forest regions of the world. Rainfall that falls on the forests of Yunnan is partially evaporated into the atmosphere, which then reinforces the northward-moving moist air currents of the monsoon. These Yunnan "forest-assisted" clouds then release their rain on Emei Shan in Sichuan, for instance, and so on, as the process continues into northern China.

The wholesale removal of forests in any region of the earth can have grave short- and long-term consequences, and China is no exception. Much of the situation is well known in scientific circles, yet it should be made clear to all of society: sound ecologically based reforestation programs are urgently needed throughout China and bordering countries to help sustain the rainfall patterns that ultimately sustain China's forests and civilization. We cannot defy the natural climatic cycles that have been a radical feature in Eurasia during the Pleistocene and Holocene. But we should acknowledge how we can gravely exacerbate these natural extremes by wholesale forest destruction and increasing industrial aerosol pollution.

> Forest is gold. If we know how to conserve it well, it will be very precious.
> Destruction of the forest will lead to serious effects on both life and productivity.
>
> —Ho Chi Minh, 1963,
> dedication ceremony for Cuc Phuong National Park, Vietnam[5]

Notes

1. A fascinating new edition of the late Kingdon Ward's *Riddle of the Tsangpo Gorges*, edited by Kenneth Cox (2001), presents lots of new geographical information and catches the stupendous scale of this amazing region.
2. Cunningham (1984) presents a fine description of Frank Meyer's dedicated exploration for economic plants in northern China on behalf of the U.S. Department of Agriculture during the early nineteenth century.
3. Readers looking for detailed information on the geography and plant exploration history of Emei Shan should read the masterly account by Roy Lancaster (1989).
4. The factual information about Chinese tiger populations in this chapter was found on the Web site for Save China's Tigers: http://www.savechinastigers.net, consulted on 15 April 2004.
5. From *Vietnam* (2003), 7th edition, by Mason Florence and Virginia Jealous (London: Lonely Planet).

Physiogeographic References

An, Z., W. Pinxian, W. Shumig, et al. 1991. Changes in the monsoon and associated environmental changes in China since the last interglacial. In T. Liu, ed. *Loess, Environment and Global Change; Series of the 8th INQUA Congress*. Beijing: Science Press.

Chang, H. 1981. The Quinghai–Xizang Plateau in relation to the vegetation of China. In *Geological and Ecological Studies of Quinghai–Xizang Plateau, Proceedings of the Symposium on Quinghai–Xizang Plateau*. Beijing: Scientific Press. 1897–1903.

Cox, E. H. M. 1945. *Plant Hunting in China*. London: Collins. 180, 183.

Cox, K., ed. *Frank Kingdon Ward's Riddle of the Tsangpo Gorges: Retracing the Epic Journey of 1924–25 in South-east Tibet*. London: Antique Collectors Club.

Cullen, J. L. 1981. Microfossil evidence for changing salinity patterns in the Bay of Bengal over the last 20,000 years. *Paleogeography, Paleoclimatology and Paleoecology* 35: 315–356.

Cunningham, I. S. 1984. *Frank N. Meyer, Plant Hunter in Asia.* Ames: Iowa State University Press.

Global Witness. 2003. *A Conflict of Interests: The Uncertain Future of Burma's Forests.* London: Global Witness.

Hopkirk, P. 1980. *Foreign Devils on the Silk Road.* London: Oxford University Press. 12.

Jarvis, D. I. 1993. Pollen evidence of changing Holocene monsoon climate in Sichuan Province, China. *Quarternary Research* 39: 325–337.

Lancaster, R. 1989. *Travels in China, a Plantsman's Paradise.* London: Antique Collectors Club. 83–85.

Lettau, H., K. Lettau, and L. C. B. Molion. 1979. Amazon's hydrologic cycle and the role of atmospheric recycling in assessing deforestation effects. *Monthly Weather Review* 107: 227–238.

Liew, P., C. Kuo, S. Y. Huang, and M.-H. Tseng. 1998. Vegetation change and terrestrial carbon storage in eastern Asia during the Last Glacial Maximum as indicated by a new pollen record for central Taiwan. *Global and Planetary Change* 16–17: 85–94.

Liew, P., C. Kuo, and M. Tseng. 1995. Vegetation of northern Taiwan during the last glacial maximum as indicated by new pollen records. *Abstracts, 14th INQUA Congress.* Berlin: INQUA. 161.

Liu, B., ed. 1996. Topography of China. In *Atlas of China* (English ed.). Beijing: China Cartographic Publishing House. 65–81.

Liu, K.-B. 1986. Pleistocene changes in vegetation and climate in China. *Conference Abstracts, American Quaternary Association.* 94.

Liu, K.-B. 1991. Quaternary vegetational history of the monsoonal regions of China. *Special Proceedings, INQUA XIII Congress, Beijing.* Berlin: INQUA. 111–113.

Liu, T. 1985. *Loess in China,* 2d ed. Beijing: China Ocean Press.

Mithen. S. 2002. *After the Ice, a Global Human History, 20,000 to 5,000 BC.* London: Weidenfeld and Nicholson.

Rabinowitz, A. 2004. A question of balance. *National Geographic* 205 (4 April): 98–117.

Salisbury, H. E. 1989. *The Black Dragon Fire.* Boston, Massachusetts: Little Brown.

Save China's Tigers. Web site http://www.savechinastigers.net.

Shupe, J. F., M. B. Hunsiker, J. F. Dorr, and O. Payne. 1995. *National Geographic Atlas of the World,* 6th ed. Washington, DC: National Geographic Society.

Winkler, M. G., and P. H. Wang. 1993. Late Quaternary vegetation and climate of China. In H. E. Wright et al., eds. *Global Climates Since the Last Glacial Maximum.* Minneapolis: University of Minnesota Press.

Ying, T., Y. Zhang, and D. A. Boufford. 1993. *The Endemic Genera of Seed Plants of China.* Beijing: Science Press.

Zhou, W., et al. 1992. Variability of monsoon climate in East Asia at the end of the last glaciation. *Quaternary Research* 39: 219–229.

Zonneveld, K. A., et al. 1997. Mechanisms forcing abrupt fluctuations of the Indian Ocean summer monsoon during the last deglaciation. *Quaternary Science Reviews* 16: 187–201.

The Issue of Bioinvasiveness

Douglas Justice

The promise of a new plant is extraordinarily attractive to gardeners. Most of us will try anything once, and it is common for gardeners to sing the praises of the very new as though newness itself is a virtue. But any account that advises gardeners on the suitability of poorly known and mostly untried plants, without first considering their destructive capacity, should be considered suspect from the start. One only needs to look around to see the devastation of native ecosystems wrought by introduced plants and animals.[1] The Australian flora has been contaminated so effectively with alien species, for example, that officials there have fashioned plant importation policies that are arguably the strictest in the world. The United States maintains a ban on commercial importation of bamboo species that was enacted in 1918; interestingly, the ban was established not to restrict bamboo importations per se, but to prevent the introduction of *Ustilago shiraiana*, a smut disease that was killing bamboos and other grasses elsewhere around the world.

Weed specialists remind us that early maturing, rapidly growing plants that produce copious amounts of easily dispersed seed are almost always the ones fingered for invasiveness, and there are many excellent examples. *Momordica charantia* (bitter melon) is a classic alien gone wild. A fast growing, vining perennial, *Momordica* produces flowers and fruits within months of germinating. Known throughout the tropics and subtropics, the plant is valued as an ornamental as well as for both its culinary and medicinal uses. These features are the same as those that helped distribute *Pueraria montana* var. *lobata* (kudzu) in the American South, a rampageous strangler and smotherer now common from the Gulf Coast north to Pennsylvania. (In fact, during the Great Depression, no less than the U.S. Soil Conservation Service promoted kudzu for erosion control.) For its many virtues (both real and imagined), bitter melon is promoted widely, and although not considered a particularly aggressive invasive, it has become naturalized throughout Micronesia and parts of the American South. As it gradually insinuates itself more and more into the landscape, building both its soil-borne bank of seeds and its ability to produce them, *Momordica* has the potential to become another serious pest.

Here in the balmy, wet Pacific Northwest, people are only just waking up to the threats of invasive species. Until recently, few believed that alien species could thrive under the climatic conditions here, which we characterize as wet and cold in the winter, with a dearth of late springtime soil moisture and an absence of significant summer heat. The ubiquity, however, of invading *Hedera helix* (English ivy), *Ilex aquifolium* (holly), and *Sorbus aucuparia* (European mountain ash) in this region contradicts the notion that slow-to-fruit perennials are not serious weeds, as well as the widespread belief that ours is a difficult climate. The Pacific Northwest can also complain of a wide variety of more typical weedy invaders, such as *Reynoutria japonica* (Japanese knotweed), *Lythrum salicaria* (purple loosestrife), and the annual *Impatiens glandulifera* (Himalayan jewelweed), each of which was introduced for its significant ornamental value.

Based on our work at the University of British Columbia Botanical Garden, we now recognize that many plants previously considered benign, especially lianas (woody climbers), are potentially pestilential and capable of invading our local woodland. We encourage lianas by training them into the branches of mature trees (30- to 50-metre conifers) and allowing them to climb without restriction, which appears to drastically increase their vigour and reproductive tendencies. As they do in the wild, almost all of the lianas eventually surmount their hosts completely, weakening them considerably in the process. Much-valued ornamentals, such as *Vitis coignetiae* (tapestry vine), have become destructively vigorous under this regime. In cultivation, these climbers are usually planted as isolated clones in restricted spaces, and in that context generally pose few problems. But in a large garden where space is practically unlimited, *V. coignetiae* can become enormous and bear many flowers. With pollen available from other individuals, the hundreds of thousands of flowers produced are frequently fertilized, producing viable seed as a consequence. Fruit-eating birds—always willing participants in this kind of study—are usually abundant, too. Such behaviour and results are well known in other regions, but are relatively novel here. Fortunately, no seedlings from our garden have appeared in the area beyond the garden, but the sheer volume of seeds significantly increases the potential for escape.

At the UBC Botanical Garden, we take seriously the threat of exotic plants becoming pests. In some cases, and often despite significant ornamental appeal, we have been compelled to remove species entirely, and to strongly recommend against their introduction as ornamentals. For example, *Hedera nepalensis, Rubus simplex, R. xanthocarpus,* and *Vitis piasezkii* var. *pagnuccii* have all proven to be extraordinarily robust and potentially invasive in the David C. Lam Asian Garden. Development of an ongoing assessment protocol for invasive potential in the garden has become a priority, as UBC continues to acquire significant numbers of exotic plants for its collections.

Some of the plants we are closely monitoring include a variety of pinnate-leaf *Sorbus* species and a small number of climbing roses, *Actinidia* and *Schisandra* species. These plants are adapted to cool, shaded forest conditions that produce fruits attractive to birds. Out of the woodland, in the UBC E. H. Lohbrunner Alpine Garden, wind-distributed seed producers such as the western American *Erigeron glaucus* have become locally weedy. Some rhizomatous plants, including the normally tender *Convolvulus althaeoides* from southern Europe, have become aggressive among the boulders in this garden. This pretty little ornamental species could be considered a threat elsewhere on similar well-drained, exposed sites. In ponds in the same area, *Nymphoides peltata* (Eurasian water fringe) is now a significant pest.

As authors of this book, we have tempered considerably our selection of the plants discussed here, to avoid recommending plants that are generally and obviously prone to invasiveness. Characteristics that contribute to a plant's ecological unsuitability on certain sites, however, may be the same qualities that confer valuable behaviour in other places. There is often a fine line between a plant's ability to naturalize and its propensity to become a pernicious weed. While we have had some difficulty in predicting the invasive potential of species for other parts of North America—particularly as many subjects do not set seed, nor grow quickly in the Vancouver area—the reader must consider local conditions and the potential that similar or related plants have shown for becoming invasive in their area.

It is inevitable that some of the recommended plants in this book will eventually prove unsuitable in some regions. Assessments sometimes take many years, as in the case of *Vitis coignetiae*, whose reproductive potential we are only now beginning to fathom.

Regarding the issue of invasiveness, it is important for all of us to observe and monitor plants in our gardens and in the wild, and to heed the warning signs of invasiveness if and when they arise.

Notes

1. See Bright (1998) for a synthesis on plant and animal bioinvasion and a discussion of "counterinvasion" techniques. Myers and Bazely (2003) examine historical, ecological, and sociological perspectives relating to invasive plants. Cronk and Fuller (2001) discuss patterns, predictability, and management options associated with invasive plants and their threats to biodiversity.

References

Bright, C. 1998. *Life Out of Bounds: Bioinvasions in a Borderless World.* Worldwatch Environmental Alert Series. New York: W. W. Norton.

Cronk, Q. C. B., and J. L. Fuller. 2001. *Plant Invaders: The Threat to Natural Ecosystems.* London: Earthscan.

Myers, J. H., and D. R. Bazely. 2003. *Ecology and Control of Introduced Plants.* Cambridge, England: Cambridge University Press.

PART 1

Perennials

BRENT HINE

Aconitum vilmorinianum Komarov
Climbing monkshood
Ranunculaceae   Plates 24–25

DISTRIBUTION: West and central China (in west Guizhou, southwest Sichuan, and central and northern Yunnan), and into Mongolia, at 2100 to 3000 m.

THIS PERENNIAL has a rare twining growth habit, and extends over the growing season to about 4 m. Most other *Aconitum* species are self-supporting on straight stems, with the exception of climbers such as *A. hemsleyanum* and *A. volubile*.

Its late start and climbing habit make it valuable in both herbaceous and mixed borders. New growth commences in late spring or early summer and proceeds slowly, gaining speed as summer heats up. In our relatively cool Pacific Northwest summers, it reaches its full potential in October, when it flowers profusely until the first hard frosts, usually in mid November. At UBC Botanical Garden (UBCBG), we grow it to advantage into a large *Rhododendron rubiginosum* and beside a path, where its sapphire blue flowers can be appreciated at close range. Different combinations with colourful deciduous shrubs would also provide glorious sights for late-season garden tours. In warmer summer climates, this plant will begin its bloom cycle much earlier, probably in mid to late August. This perennial is long-blooming. As with all *Aconitum* (monkshood) species, all parts are poisonous. This well-mannered, richly flowered, extraordinary (and valuable) plant is nearly unknown to North American gardeners.

HARDINESS: Zones 5–9. The species is probably hardier than indicated in the literature to date, to Zone 5 at least. Native to a wide area in central Asia, it grows mostly at subalpine elevations in shrubby thickets.

CULTIVATION: A woodland-edge relationship suits the plant best. Here, its feet are in the shade, there is a suitable scaffold for it to grow upon (though it can also scramble over the ground), and it will climb toward sunnier reaches. It best likes a moisture-retentive soil, which will create vigorous growth. In climates colder than those of the Pacific Northwest, any lasting snow cover will increase hardiness. Otherwise, planting deeper than usual will help, but will delay bloom somewhat.

PROPAGATION: As with its close relative, var. *altifidum*, this perennial is easily divided by pulling apart the clump at the new individual crowns. Seed is not produced in our region's cool-summer climate, as the plant blooms late.

Aconitum vilmorinianum var. *altifidum* W. T. Wang
Ranunculaceae   Plates 26–27

DISTRIBUTION: China, in Sichuan.

THE VAST REACHES of China have been fruitful ground for producing *Aconitum* species. Among them, this special plant belongs in the rare category of semiscandent perennials. Its varietal name indicates that it is a "faithfully lofty" climber, but even so, it does not achieve the towering growth characteristic of the species. Instead, it develops like something between a bush and a climber. As the growing season progresses, its multiple stems elongate and bend over under their own weight. This woodland-edge perennial therefore should be cultivated with support from sturdier, woody branches. As its stems wind onto scaffolds, it clothes them with its deeply divided leaves, and the plant ultimately assumes a mature height of about 1.5 m. By midsummer, it produces twilight blue-purple flowers with a pearly cast that offsets its striking dark stamens. This colouration seems to suit its desire for partially shaded habitats, and contributes to an air of mystery. The UBCBG plant grows next to a large, straw-coloured boulder with adjacent ferns and toad lilies (*Tricyrtis* spp.), which complement the foliage and flowers.

In our experience at UBCBG, these plants are tough. They are not bothered by any notable insect or disease problems in the garden. Quite surprisingly (or perhaps not, for Chinese perennials on the whole are mostly robust here), they have adapted well to our lowland maritime climate without slackness in performance, and are near to perfectly behaved and happy plants. As with all *Aconitum*, these plants are poisonous and should be placed thoughtfully in the garden.

These perennials should adapt easily to garden cultivation throughout most of temperate North America, where their preferred habitats of partial shade and moisture are available to them. For their unique habit and exquisite flowers, they might be tried by gardeners desiring something unusual.

HARDINESS: Zones 5–9.

CULTIVATION: As with other *Aconitum* species, this one prefers soil that retains moisture throughout the growing season. After initial planting, its tuberous roots will settle down quickly and will form a shovel-sized clump in a couple of years. After that, an annual application of well-rotted manure will provide greater health and copious midsummer bloom. Staking is optional, but unnecessary when the plant is positioned well in the garden. Division of plants is best performed in early spring.

PROPAGATION: Unlike similar aconitums, this one blooms early enough in the season that plenty of seed can mature before season's end. As well, clumps are robust, providing plenty of material for division. When dug, vertical tuber-like roots (technically, each is a caudex) are revealed. These break away from the parent to form new crowns, and so are easily separated and moved to new garden homes.

Actaea japonica Thunberg var. *japonica*
Maple leaf fairy candle
syn. *Cimicifuga japonica* (Thunberg) Sprengel
Ranunculaceae   Plates 28–29

DISTRIBUTION: Japan, Korea, and China, over a wide area, at 800 to 2600 m.

THE GENUS *Cimicifuga* is commonly known to North American gardeners as bugbane, referring to the plant's alleged ability to repel insects. *Actaea*, another ranunculoid genus, is known widely as baneberry due to its poisonous fruit. Both curious and beautiful, the perennial members of this genus are often sought as refined and easy-care plants. As you can tell from the synonymy, this plant is both a source of ongoing disagreement and a welcome addition to an aristocratic group. In late summer and early fall, whole mountainsides in Japan and China are adorned with spikes of its pearl-budded white flowers; rising to 1 m, the flowering stems are unabashedly attractive. The bold leaves, remarkable as well, are prominently veined and patterned in the shape of a maple leaf, giving this species its common name.

Most current authorities have sunk the genus *Cimicifuga* into *Actaea*. At the same time, *Flora of China* (Wu and Raven 2001) has kept its original classification of *Cimicifuga*; the two genera are separated partly on the basis of their fruits. While Chinese actaeas have a berry, *Cimicifuga* employs a dehiscent (drying, splitting) follicle arrangement. Other authorities do not find this and other differences sufficiently distinctive, and so the debate continues between the "lumpers and splitters." In any case, as a group they are ancient plants, with their archaic centre of distribution in East Asia. The taxonomists will eventually sort out the phylogeny; the ability to grow and appreciate the plants is perfectly satisfying for the rest of us. At UBCBG, we accept the most recent generic classification of this species in *Actaea*.

Gardeners who appreciate the combination of exotic leaves and sophisticated floral spikes will find this perennial made to order. Ultimately, the plant is both useful for the practical gardener and designer and appreciated by the collector of rarities. It also has consistent, firm hardiness, so has wide application in North America.

HARDINESS: Zones 4b–9.

CULTIVATION: Preferring partially forested hills, this perennial appreciates meadow situations—bright sun, not too hot, and rich soil. It is ideal for use in perennial borders and woodland-edge gardens with moderate afternoon shade. It spreads by rhizomes, but only modestly.

PROPAGATION: This fibrous-rooted species is an easily managed, if slow-to-increase, perennial. Its handsome character can be especially appreciated in a group planting. Since it blooms late in the summer, it should be divided in mid-spring, after soil has warmed and growth has begun.

Artemisia rutifolia Sprengel
Asteraceae   Plate 30

DISTRIBUTION: Tajikistan, in the eastern Pamir Range, between 3500 and 4200 m.

THIS CURIOUS and compact herbaceous perennial hails from the farthest flung regions of western Asia, where it occupies some of the most inaccessible mountain haunts imaginable. Think of the upper reaches of the Pamir Mountains, lying north of Afghanistan, just west of westernmost China, at altitudes where solar radiation, cold, and wind are extreme in even the most civilized of gardens. When I was introduced to this perennial 10 years ago in the more agreeable context of the UBC Botanical Garden, I was intrigued because its native habitat was described as "cryophytic (growing with snow and ice) sagebrush deserts." I was stunned, then, to find that the plant has responded so favourably to its climatic opposite in the rainy, mild Pacific Northwest environment and, moreover, done so with vigour.

Having already proven its adaptability to cultivation in our garden, this species possesses special qualities that will also appeal to many gardeners. Leafing out from its slender woody stems very early, the plant soon attracts attention. Its finely dissected leaves are soft greyish green and form into a billowing filigree that rivals the best of the genus. As summer progresses and rainfall on the West Coast decreases, the plant gains a more intense silver cast, as long as irrigation is restrained. The species name *rutifolia* describes the divided leaf structure, which resembles rue, but evokes the unique and subtle beauty of desert plants in general. The first time I touched the leaf, I was astonished to discover that it possesses an intense wormwood fragrance, yet it conjures something more. I was suddenly in contact with childhood memories of poking around in dusty old garages, and coming across, among other things, old half-used cans of paint. The essential oils (turpenes) in this artemisia are a surefire stimulus to the senses.

During June, *Artemisia rutifolia* develops numerous panicles of rather inconspicuous flowers that reach to about 30 cm. Since the flowers are not outstanding, removing them after a while brings focus back to the plant's special foliar qualities. While a few artemisias, such as *A. schmidtiana* 'Nana' and *A. stelleriana*, are excellent garden plants, many more are best left to be admired in their desert and mountain homes. This one, however, is also a distinct artemisia that thrives right where many can appreciate it—in our low-maintenance garden.

HARDINESS: Zones 3–9.

CULTIVATION: This perennial is quite at home in the gravelly soil of the E. H. Lohbrunner Alpine Garden, where it appreciates free drainage and the warm, southwest-sloping site. As with most artemisias, this one benefits from a good clipping, both in early spring and after blooming, to encourage denser growth and foliage display. The plant will appeal especially to those with seasonally drier gardens.

PROPAGATION: The best method for propagating this perennial is to take softwood cuttings during early summer. These root quickly and will produce viable young plants ready for planting in early fall. This perennial does not produce seedlings in cultivation: however, it sends up suckers around the parent plant, which allows for careful division, best done before summer arrives.

Calamagrostis hakonensis Franchet & Savatier
Poaceae   Plate 31

DISTRIBUTION: Throughout Japan and the South Kurile Islands northeast of Hokkaido, and China, in low mountains.

MANY GRASS-LOVING gardeners have heard of *Hakonechloa macra* (Hakone grass), and many have grown its famous cultivar 'Aureola'. If this applies, you should not only see, but grow *Calamagrostis hakonensis*. In cultivation it stands alongside that other grass and distinguishes itself very well indeed. It can grow among other perennials without creeping in and muddling things up, as Hakone grass will do when left on its own for long. It is a calmly spreading species in contrast to many others of its kin that are also known for their beauty but can all too easily spread aggressively on the rich diet of a formalized garden setting.

This small perennial, growing only to about 60 cm, is closer to 40 cm in our garden. Its light and airy appearance is a refreshing change from most other cultivated *Calamagrostis*, whose appearance can dominate in the garden. Its arching, medium-weight leaves are polished apple green. The flower heads, which appear after midsummer, are refined yet substantial and charming. When the cooler nights of fall return, the leaves lighten to a warm yellow, before turning buff to end the growing season.

I first cultivated this plant and then had the good fortune to experience it in the wild, all within a short time. While hiking in the mountains at Hakone, Japan, I suddenly became aware of its presence in that unique landscape. In the vicinity of several ancient *Enkianthus*, growing among broken lava rock and a foul sulphurous odour emitting from nearby volcanic vents, there it was, thriving in abundance. Seeing it in its chaotic natural habitat gave me a heightened appreciation for the plant's character. Back in the E. H. Lohbrunner Alpine Garden, it grows contentedly in a moist, sunny bed with autumn-flowering Asian gentians and *Salix purpurea* 'Pendula', a most complementary combination. It fits easily into the general cultivated landscape, too. In North American gardens, its graceful, controlled habit suggests wide use both in more formal city gardens and relaxed country gardens.

HARDINESS: Zones 5–9.

CULTIVATION: From mountain woods and open slopes, it is especially well suited to sunny and well-drained sites where there is moisture available at the root. It is easily divided if placed adjacent to less assertive neighbours and will settle down quickly again.

PROPAGATION: Every year a few seedlings are found around the parents. Established clumps are not too big or troublesome (some *Miscanthus* come to mind) to divide. Do this during spring, just before plants begin active growth. Separate them with a sharp knife or pair of secateurs and patiently tease them apart. Each division will have its own roots and will take a full season to get established in the garden.

Clematis ranunculoides Franchet var. *ranunculoides*
Ranunculaceae   Plate 32

DISTRIBUTION: Southwest China, in northwest Guangxi, western Guizhou, southwest Sichuan, and Yunnan, at 500 to 3000 m, growing in open grass, scrub, and forests, sometimes along streams.

THIS LITTLE-KNOWN clematis, placed in the section *Campanella,* is among more than 40 other species native to China. It is described as a variable but generally shrubby species. One of these, a scrambling or climbing form, has been growing at UBCBG since 1992. It has many exciting virtues.

Clematis ranunculoides delights us during late fall, when most other herbaceous perennials have faded for the season. Its dark green, conspicuously hairy stems elongate to about 3 m, snaking through nearby shrubbery. Only by September's end, when I have nearly given up hope, do the year's first flower buds finally appear. As fall weather cools considerably, one wonders whether there will be a show of flowers before frost. Indeed, some years, blossoms have been cut short by an early blast of winter. Usually, however, by mid October these buds unfurl into many small, charming, rosy red to violet, campanulate bells with recurved tepals. The flowers are borne in the axils of the much-divided leaves. (This division of the leaves is responsible for the species name, which refers to the plant's resemblance to the herb ranunculus.) The flowers are arranged in small groups of 2 to 3 or more, mostly near or at the terminal end of the plant.

Without a doubt, this whole blooming affair would begin earlier (and probably last longer) in warmer climates than the Pacific Northwest. There are indeed many Asian clematis that provide us with late spring and early summer inspiration, whereas some of the fall bloomers (such as *Clematis orientalis* group) grow to enormous proportions, beyond the scope of most home gardens. *Clematis ranunculoides,* however, is a modest species that asks only for a warm place to scramble, and will provide a memorable late show.

HARDINESS: Zones 7–10.

CULTIVATION: Christopher Grey-Wilson (2000, 134) states in *Clematis, the Genus* that the plant's chief enemy is not extreme cold but "excessive winter wet." Our plant is placed next to an ancient, massive tree stump and a similarly scaled *Gaultheria shallon,* so its stems can gain the height they crave. Growth is also enhanced by its placement on the south side of the stump, giving it warmth during short fall days. It clearly prefers a drier habitat than many other species, at least in our Pacific Northwest climate. In China, however, these plants grow in a wide variety of habitats, from open meadows (shrubby types) to forest-edge situations (scramblers), yet common to all is plenty of light. Because of the wide ecotypical variance in this species, its ability to thrive depends on the site where a plant is grown.

PROPAGATION: During late summer, semiripe cuttings may be taken. I would use divisions of the plant, to begin with a better root system. Because bloom is achieved so late in the season, seed propagation is not a viable option in the Pacific Northwest.

Delphinium ceratophorum Franchet (CLD 0893)
Ranunculaceae   Plate 33

DISTRIBUTION: China, in Yunnan, between 2800 and 4200 m, among shrubs and trees in open situations.

LEADING THE CHARGE away from high-maintenance gardens, many knowledgeable gardeners seek showy, easy-going (and growing) perennials, especially "wild" perennials, or those that will take care of themselves within the bounds of cultivation. Full-season performance remains key, and that is where the best of the Asian *Delphinium* species come to the fore. This plant is just becoming known in gardening circles in North America, so far mostly from expedition seed collections.

The species name *ceratophorum* describes the horned appearance of the flower spurs, but does not convey the beauty of this plant both in or out of flower. Once established, it pushes out of the ground from multiple stems in March. Its growth habit forms a miniature thicket of light green stems bearing bold, many-lobed, dissected leaves that offer a distinctive ornamental quality. Eventually, stem height is about 75 cm, and because the stems are soft, that is not a problem. This woodland-edge perennial is not a big-flowered, pumped-up hybrid, so it does not need to be staked.

The end of June usually heralds the start of summer in the Pacific Northwest, and that is when the mass exhibition of bright blue flowers begins. Starting from the top down, they are borne copiously from the leaf axils along the entire upper half of each stem. Individually, the flowers are not large, but the overall effect is stunning. At UBCBG, the plant is sited in a bed rich with humus both natural and applied and receives morning sun. In this mimicry of its natural habitat, its needs are satisfied, leading to years of carefree enjoyment.

While this plant is endemic to the moderately high Yunnan plateau at least 800 km from the Pacific Ocean and far south of our northern shore, it has proven a winner for us. Different clonal selections also suggest that it will be hardy for cooler locations within North America. *Delphinium ceratophorum* is a viable alternative to many of the popular but attention-needy garden hybrids.

HARDINESS: Zones 5–9.

CULTIVATION: Copious humus is needed, to retain moisture during the growing season and achieve robust growth. For best results, plant away from the sun's hottest aspect.

PROPAGATION: Due to the reliable yearly mass of flowers, a richness of seed is also produced. Exchange some with your dedicated gardening friends. In a few seasons, the clump will be approximately 1 m across and ready for division as growth commences above ground in spring.

Dendranthema mongolicum (Ling) Tzvelev
Mongolian daisy
syn. *Chrysanthemum mongolicum* Ling   Plate 34

Chrysanthemum yesoense Maekawa
Yezo daisy
syn. *Dendranthema yesoense* (Maekawa) D. J. N. Hind   Plate 35

Asteraceae

DISTRIBUTION: Carpathian Mountains of Ukraine in western Asia, through Russia, China, Korea, and Japan, on the ocean.

THESE CHARMING perennials are part of an enormous group of low-growing, undemanding, and floriferous daisies. Many were in cultivation centuries ago, recognized for their simple beauty and medicinal attributes. Most species derive from continental Asia, from such an extraordinary range of geographical regions that implies their suitability for myriad garden settings. And indeed, this genus has great hardiness and vigour. Natural habitats are also widely disparate, from rock-strewn boulder fields and mountain meadows to sandy beaches. Daisies have become indispensable garden plants. Whether for use in rock (alpine) gardens (as at UBCBG), city park wildflower meadows (finding pride of place in Europe), or herbaceous perennial border culture—whether from cold-temperate to subtropical climates—they will thrive.

Dendranthema mongolicum begins its flower cycle in late spring, with a soft pink explosion of flowers as potent as any primula. The other "mums" effectively take the late-season garden stage during a time when little else is in flower. Mongolian daisy spreads slowly by rhizomes to form a solid, dark green mat with a crunchy cut-leaf texture that warms to yellow and orange tones in fall. With age, the plant extends to 1 m or more. The slow rhizomatous spread of the perennial can go on indefinitely, but pieces are easily separated.

Blooms of *Chrysanthemum yesoense* arrive a bit later than those of *Dendranthema mongolicum*, in early summer, on short stems of not more than 30 cm. Flower colour is pure white. Flowering of these perennials, whether achieved early or very late in the season, is valued for its volume and longevity.

A bonus feature of these survivor plants is their modest water needs. Yes, in the end, they are just daisies, and yet, when autumn comes and gardeners are searching for lovely late-bloomers, *Chrysanthemum yesoense* gives a grand performance.

HARDINESS: *Dendranthema mongolicum*, Zones 4–9; *Chrysanthemum yesoense*, Zones 3–9.

CULTIVATION: Plant in any fertile, exposed, sunny site.

PROPAGATION: Divide established clumps, whether rhizomatous or stoloniferous, with a sharp spade. Place in moistened soil with a sprinkling of 6–8–6 fertilizer. The result? Large, thriving colonies of plants.

Dendranthema zawadskii (Herbich) Tzvelev
Siberian daisy
syn. *Chrysanthemum zawadskii* Herbich
Asteraceae

DISTRIBUTION: Carpathian Mountains of Ukraine in western Asia, through Russia, China, Korea, and Japan, on the ocean.

TALLER THAN *Dendranthema mongolicum* or *Chrysanthemum yesoense*, *D. zawadskii* is more loose and open in habit, at twice the height. Its flowers are among the last in the garden to greet the waning fall sunrays face on, as obedient sunflowers do. Flower colour is dark pink.

Like *Dendranthema mongolicum* and *Chrysanthemum yesoense*, *D. zawadskii* has modest water needs and gives a grand performance in autumn.

HARDINESS: Zones 5–9.

CULTIVATION: Plant in any fertile, exposed, sunny site.

PROPAGATION: Divide established clumps, whether rhizomatous or stoloniferous, with a sharp spade. Place in moistened soil with a sprinkling of 6–8–6 fertilizer. The result? Large, thriving colonies of plants.

Ellisiophyllum pinnatum (Bentham) Makino
Scrophulariaceae   Plate 36

DISTRIBUTION: India to Japan, and Taiwan to eastern Papua New Guinea.

THIS CHARMING, floriferous Asian perennial ground cover is only beginning to make a well-deserved name for itself in North American gardens. It is a welcome addition to the semiwild and woodland garden at UBCBG. Right at home here, it has proved both hardy and attractive throughout the year. Its best features are present when placed along path edges, where it can be viewed close up and allowed to wander freely. Travelling at a moderate pace for us, it has prospered in the humic soil of the David C. Lam Asian Garden with moisture at the root during most of the year. In such conditions, it creates a wealth of small and sparkling, pure white, 5-petaled flowers with yellow eyes, offered over an extremely long season, beginning as early as January. Flowering peaks in early summer, yet continues until sometime in November, when sharper frosts arrive in the Pacific Northwest. An added bonus to this display is the "self-cleaning" nature of the flowers, which describes their habit of dropping away cleanly from the plant when they have faded.

The specific epithet *pinnatum* refers to the plant's appealing leaves, and indeed the softly hairy plumes are another facet of its neat display. Near-relatives *Bacopa* and *Sutera* also provide us with carpeting perennials, and these are used increasingly in hanging baskets. *Ellisiophyllum*, with its robust behaviour, attractive flowers, and long bloom period, could also be utilized in this way, further increasing its horticultural potential. Whether hanging in our gardens and seen close at hand or spreading out under deciduous tree cover, this multi-use plant is attractive for all the right reasons.

HARDINESS: Zones 6–9.

CULTIVATION: *Ellisiophyllum pinnatum* is very adaptable in the garden. Although it enjoys a semishaded site, it will also tolerate full sun, given reasonable soil moisture at all times, and will submit to heavy shade, but then flower production will be diminished. Soil should be fertile, such as in a woodland garden.

PROPAGATION: Rooting as it spreads, this perennial is "the little engine that could." Left alone in leafy, moisture-retentive soil, it will fill an area at a moderate pace. Divide plants in the spring, and after a couple of growing seasons, there should be plenty of plants for you and your friends.

Epimedium brevicornu f. *rotundatum* Maximowicz (Og. 82010)
Barrenwort
Berberidaceae   Plate 37

DISTRIBUTION: China, in Gansu, Guizhou, Shanxi, Shaanxi, and Sichuan, at 800 to 2100 m, in open yet shady forests and clearings.

IF THE THREE new epimediums described in this book are destined to enrich woodland gardens throughout North America, it will be because they have several good points. They are easy to manage and have adapted flawlessly to culture at UBCBG. They are beautiful as well as practical for use as ground cover throughout the growing season, not for a mere few weeks while in bloom. Their foliage is delightful while unpalatable to most garden pests. They carry with them one of the most desirable traits of the genus, toughness. Once established, these perennials can tolerate the most challenging of temperate garden sites—dry shade. And they have cachet that will appeal to the collector of valuable herbaceous perennials. We are pleased to introduce these epimediums to a wider North American audience.

Epimedium brevicornu f. *rotundatum* is a more delicate-looking plant than many of its genus. Compared with the typical species, it is a half-sized, 30-cm-tall, round-leaf form. The species name *brevicornu* refers to the plant's short, or brief, petals. Due to the plant's origin on damp limestone cliffs, most experts assumed that the plant would be a challenge to cultivate, but experience at UBCBG has shown the opposite. In the acidic woodland conditions of the David C. Lam Asian Garden, it has taken to cultivation very well, spreading by slow-creeping rhizomes, the means of increase common to all epimediums. Moisture-retentive soil has enabled it to continue growth and hold its attractive form during the long growing season. It is especially a joy to behold in spring, when it bears its profuse white flowers with yellow centers that are held aloft on wiry stems over the fresh green foliage. The leaves are dainty, with minutely spined margins. Unlike the leaves of some other species, the leaves are not at all coarse; they are reminiscent of the leaves of the related genus *Vancouveria*.

HARDINESS: Zones 5–9.

CULTIVATION: *Epimedium brevicornu* f. *rotundatum* will create sizeable colonies within 2 or 3 years, if plants are provided with reasonable nutrition in the form of humus or leaf litter. Seasonal moisture is important for rhizome development. In spite of the locally summer-dry climate (opposite to that of many epimediums), this botanical form has adapted. If it becomes semideciduous in cooler North American climates, remove the previous year's leaves before new growth begins. It socializes well with such companions as mixed ferns, *Reineckia carnea*, or *Rohdea japonica*.

PROPAGATION: As a rule, epimediums are not only social, they are promiscuous. Seed is therefore not reliably true if more than one taxon is planted in a group. The seeds are also very tiny and a bit troublesome to collect. Divisions are preferred, and within a couple of years, there should be enough to spread around. Divide in early spring as growth begins, or wait until flowering is finished, if soil moisture is consistently available.

Epimedium leptorrhizum W. T. Stearn (Og. 91011)
Berberidaceae   Plate 38

DISTRIBUTION: China, in Gansu, Guizhou, Shanxi, Shaanxi, and Sichuan, at 800 to 2100 m, in open yet shady forests and clearings.

THE SPECIFIC epithet *leptorrhizum* describes the plant's slender rhizome. Like others in this genus, the plant looks best during spring, when new leaves are expanding in groups of 3 leaflets and the graceful flowers appear. The early leaves are suffused with a rich red wine colour, a beautiful sight to behold. The immature leaf colouration lasts as long as spring in the Pacific Northwest—sometimes several months. Meanwhile, the flowers hang close-grouped at the terminals of the camouflaged stems, and are displayed in captivating shades of rosy pink to mauve. Similar to other epimediums, the mature plants are tough, reminding us of their berberidaceous cousins, mahonias. They carry themselves with sophistication, while managing their shady summer drought conditions without difficulty.

HARDINESS: Zones 6b–9.

CULTIVATION: Like *Epimedium brevicornu* f. *rotundatum*, *E. leptorrhizum* will create sizeable colonies within 2 or 3 years, if plants are provided with reasonable nutrition in the form of humus or leaf litter. Seasonal moisture is important for rhizome development. In spite of the locally summer-dry climate (opposite to that of many epimediums), this species has adapted. If it becomes semideciduous in cooler North American climates, remove the previous year's leaves before new growth begins. It socializes well with such companions as mixed ferns, *Reineckia carnea*, or *Rohdea japonica*.

PROPAGATION: As a rule, epimediums are not only social, they are promiscuous. Seed is therefore not reliably true if more than one taxon is planted in a group. The seeds are also very tiny and a bit troublesome to collect. Divisions are preferred, and within a couple of years, there should be enough to spread around. Divide in early spring as growth begins, or wait until flowering is finished, if soil moisture is consistently available.

Epimedium ogisui W. T. Stearn
Berberidaceae   Plate 39

DISTRIBUTION: China, in Gansu, Guizhou, Shanxi, Shaanxi, and Sichuan, at 800 to 2100 m, in open yet shady forests and clearings.

A DELIGHTFUL ornamental perennial, *Epimedium ogisui* is named in honour of Mikinori Ogisu, by whose determined efforts several Chinese *Epimedium* species have been introduced to Western cultivation and study. This one is horticulturally exceptional because of its compact yet robust habit, its enticing flowers, and its compliant culture. At UBCBG, under mixed cover of evergreen and deciduous trees, it thrives with the same ease of care for which the rest of the genus is noted. It is deciduous in the Pacific Northwest climate, but does not stay so for very long. By March, its pristine white flowers are already abundant. They are large (about 2.5 cm diameter), and the curved petal spurs that many *Epimedium* species display are showily evident. It is no wonder that the Japanese, with their seafaring history, have named the genus *ikari-sou*, or anchor plant. I much prefer this name to the Western colloquial name barrenwort, because that unfortunate label contradicts the graceful qualities these plants display for most of the year.

HARDINESS: Zones 7–9.

CULTIVATION: Like *Epimedium brevicornu* f. *rotundatum* and *E. leptorrhizum*, *E. ogisui* will create sizeable colonies within 2 or 3 years, if plants are provided with reasonable nutrition in the form of humus or leaf litter. Seasonal moisture is important for rhizome development. In spite of the locally summer-dry climate (opposite to that of many epimediums), this species has adapted. If it becomes semideciduous in cooler North American climates, remove the previous year's leaves before new growth begins. It socializes well with such companions as mixed ferns, *Reineckia carnea*, or *Rohdea japonica*.

PROPAGATION: As a rule, epimediums are not only social, they are promiscuous. Seed is therefore not reliably true if more than one taxon is planted in a group. The seeds are also very tiny and a bit troublesome to collect. Divisions are preferred, and within a couple of years, there should be enough to spread around. Divide in early spring as growth begins, or wait until flowering is finished, if soil moisture is consistently available.

Filipendula kamtschatica Franchet
Kamchatka meadowsweet
Rosaceae   Plate 40

DISTRIBUTION: From eastern and northern Russia, Japan, and into coastal Alaska.

AS ITS SPECIES name implies, *Filipendula kamtschatica* is a very hardy perennial that has its centre of distribution in some of Asia's (and the planet's) most remote regions. It is a giant, as perennials go, but in cultivation this meadowsweet is surprisingly understated and easily managed. At UBCBG, its bold foliage and large, frothy blooms are two inviting features that I look forward to each year. Its preferred habitat is along forest edges in mountains, and in high meadows. A must for its well-being is moisture at the root, and in fact it is often found alongside rushing streams. When luxuriating in a humid environment, it can grow to an amazing 3 m. It is no wonder that it has succeeded so admirably, growing at a creekside in our garden.

The foliage is handsome throughout the season, and its coarseness befits its origins. The large leaves (up to 30 cm) are widely palmate and rich green. Not surprising, they look even more appealing after a good rain, glistening and dripping, heavy with moisture. Because of the plant's generous proportions, it carries an almost tropical look, especially when viewed from a distance. By midsummer,

Plate 24. *Aconitum vilmorinianum* (climbing monkshood), a handsome, ambitious perennial. Photo by Judy Newton

Plate 25. *Aconitum vilmorinianum* reaching for light through *Rhododendron rubiginosum*. Photo by Judy Newton

Plate 26. *Aconitum vilmorinianum* var. *altifidum*. Photo by June West

Plate 27.
The blue, hooded
flowers of *Aconitum
vilmorinianum*
var. *altifidum.*
Photo by June West

Plate 28. The bold foliage of *Actaea japonica*
var. *japonica* (maple leaf fairy candle). Photo
by June West

Plate 29. A branched flowering stem of *Actaea
japonica* var. *japonica.* Photo by June West

Plate 30. *Artemisia rutifolia*'s delicate, silver-green foliage in an alpine garden habitat. Photo by June West

Plate 31. Elegant, informal *Calamagrostis hakonensis*. Photo by June West

Plate 32. Late blooms of *Clematis ranunculoides* var. *ranunculoides* among *Miscanthus* foliage. Photo by June West

Plate 33. *Delphinium ceratophorum* (CLD 0893) displays its abundance of azure blue summer blooms. Photo by June West

Plate 34. *Dendranthema mongolicum* on a sunny bank in an alpine garden setting. Photo by June West

Plate 35. *Chrysanthemum yesoense*, a short, sweet perennial for the rock garden. Photo by June West

Plate 36.
Long-flowering
*Ellisiophyllum
pinnatum*, a refined
ground cover.
Photo by June West

Plate 37. The flowers of *Epimedium
brevicornu* f. *rotundatum* (Og. 82010), a
woodland delight. Photo by June West

Plate 38. Large, smoky lilac flowers of
Epimedium leptorrhizum (Og. 91011). Photo
by June West

Plate 39. The pure white, substantial flowers of *Epimedium ogisui*. Photo by June West

Plate 40. The graceful flowers of *Filipendula kamtschatica* (Kamchatka meadowsweet), floating above the handsome foliage. Photo by June West

Plate 41. The compact habit of *Geranium shikokianum* var. *yoshiianum.* Photo by June West

Plate 42. *Geranium shikokianum* var. *yoshiianum*, a floriferous, long-blooming perennial. Photo by Brent Hine

Plate 43. The arresting flowers of *Geranium sinense*. Photo by June West

Plate 44. The flowers of *Geranium soboliferum*, of a deeply saturated hue. Photo by June West

Plate 45. Flowers of *Hedychium forrestii* var *forrestii*, creating summer fireworks in the perennial border. Photo by June West

Plate 46. The tropical-looking flower spikes of hardy *Hedychium yunnanense*. Photo by June West

Plate 47. *Impatiens omeiana* (Mount Omei balsam) proudly displaying bold, veined foliage. Photo by Judy Newton

Plate 48. A stem of *Impatiens omeiana*, bearing its tawny yellow flowers that resemble small goldfish. Photo by Judy Newton

Plate 49. Charming pink blooms of *Incarvillea compacta*. Photo by June West

Plate 50. Stems and flowers of *Lilium taliense*. Photo by June West

Above right:
Plate 51. *Lilium taliense*, a noble wildflower. Photo by June West

Right:
Plate 52. Swirling pink cups of *Linum hypericifolium* flowers. Photo by June West

Plate 53. Joyful late-season flowers of *Nipponanthemum nipponicum*. Photo by Brent Hine

Plate 54. The unique architecture of *Paris fargesii* var. *petiolata*, a standout in the garden. Photo by June West

Plate 55. The striking inflorescence of *Paris fargesii* var. *petiolata*, rising well above the foliage. Photo by June West

Plate 56. *Persicaria tenuicaulis*, a charming, early blooming woodland perennial. Photo by June West

Plate 57. The long-lasting bloom of *Persicaria vacciniifolia* in late summer. Photo by June West

Plate 58. *Phlomis cashmeriana* (Kashmir sage), at its zenith in the E. H. Lohbrunner Alpine Garden at UBCBG. Photo by June West

Plate 59. The blooms of *Phytolacca clavigera* (Chinese pokeweed), like a sugar pink confection. Photo by June West

Plate 60. Fruits of *Phytolacca clavigera* glow with jewel-like magnetism. Photo by June West

Plate 61. *Potentilla cuneata* showing off its bold yellow flowers and ground-covering habit. Photo by June West

Plate 62. The nodding, pale pink flowers of *Ranzania japonica.* Photo by June West

Plate 63. *Rehmannia glutinosa* (Chinese foxglove), blooming generously in cultivation. Photo courtesy of Siskiyou Rare Plant Nursery

Plate 64. Compact, robust *Roscoea tibetica* in full flower. Photo by June West

Left: Plate 65. Elegant *Salvia koyamae* (Japanese yellow sage), at home at the forest's edge. Photo by June West

Below left: Plate 66. The striking leaves and flowers of *Saxifraga stolonifera* 'Cuscutiformis'. Photo by Brent Hine

Plate 67. *Senecio cannabifolius* (Aleutian ragwort) leaf. Photo by June West

Plate 68. A massed July display of
Strobilanthes attenuata. Photo by Brent Hine

Plate 69. The richly coloured flowers of
Strobilanthes attenuata. Photo by Brent Hine

Plate 70. The sparkling *Veronica ornata* growing in the E. H. Lohbrunner Alpine Garden.
Photo by June West

pure white flowers (some selections with pale pink flowers are known) top the plant in large, lacey corymbs. These blossoms are probably the inspiration for the Japanese common name, roughly translated "ghost spiraea," and the plant is in the same family as that genus. When flowering commences, the erect stems sag somewhat, giving the impression that the plant is relaxing after its prodigious growth, but it remains well clothed in leaf.

Planted singly or in groups, as a specimen or in combination with other similar perennials and shrubs, this adaptable, pest-free (and insect-friendly), valuable Asian perennial deserves much wider use in North American gardens. It is the best kind of perennial, needing a happy minimalist approach: plant, stand back, and admire.

HARDINESS: Zones 3–9.

CULTIVATION: It thrives in humus-rich soil, in preferably part- to full-sun aspect. Siting it near any kind of informal water feature will set off its bold foliage. In hot summer areas, soil should be kept consistently moist. Judging by its character, it is best suited for the bog or wild garden.

PROPAGATION: While seed can be collected in great quantity in early fall, the rootstock can be easily split during the spring. In damp soil, root systems take on large proportions in a short time, providing plenty of raw material. This perennial shows best en masse in large-scale gardens.

Geranium shikokianum var. *yoshiianum* (Koidzumi) Hara
Geraniaceae   Plates 41–42

DISTRIBUTION: Yakushima Island in southern Japan, at high elevations.

THIS PETITE addition to the more than 300 *Geranium* species lends itself to use in uncommon gardens. It has come into cultivation from the idyllic plant paradise of Yakushima Island (south of Kyushu); many others from there have proven their worth in cultivation worldwide. The species itself, *G. shikokianum* (from Shikoku Island, slightly farther north), is at least twice the size of this variety. Our specimens at UBCBG are only 15 cm tall, so they are nearly ground cover in habit. Evolution in high mountains has also endowed the plant with sufficient hardiness for use in North America's lowland gardens; however, because this mountain plant is normally protected under snow cover during winter, its potential performance can be affected in marginal garden climates in North America.

The leaves are acutely 5-lobed, faintly yellow marbled, and softly hairy to the touch. Flowers are bright white, about 2.5 cm across, and complemented with lace-patterned purple veins. Plants begin flowering in midsummer and continue unabated until nights turn cool in late September. If positioned to take advantage of full (but not too hot) sun and receiving steady moisture at the root, this geranium continues to produce flowers up to the first hard frost. Coming from the mountainside, it also relishes a somewhat mineral soil, yet it has easily adapted to a richer diet at UBCBG while remaining in character. It slowly expands its territory both by tiny rhizomes and gentle seeding, retaining its dwarf stature and yet never getting out of hand.

Hardy geraniums have proved to be deservedly popular in cultivation; they give a lot in return for little investment of effort. This plant will be practical for use in city gardens, where its adaptability and long floral display set it apart from most of the short-lived perennials with short bloom periods. This true dwarf variety has proved to be very gardenworthy. It will easily please in a wide range of North American gardens.

HARDINESS: Zones 6–9.

CULTIVATION: This special variety needs no cossetting; it will prosper in full sun to partly shady aspects in well-drained yet moisture-retentive soil, as a member of a subgroup of meadow-inhabiting geraniums.

PROPAGATION: As with other *Geranium* species, this one produces seed in collectable amounts, a preferable way to propagate the plant. Cuttings can be taken in early summer, though, which provides relatively successful results. Division is another option, but because of the diminutive crown, it must be done with care. This perennial creates a few seedlings, which should be watched for as the growing season progresses.

Geranium sinense Knuth
Geraniaceae   Plate 43

DISTRIBUTION: Southwestern China, in woodland edges and other partially shaded locations.

JUST WHEN you think you have a handle on knowing and growing the horticulturally important members of the genus *Geranium*, along comes this unusually flowered species from southwest China. One of five Asian species with reflexed petals, it is currently found only in a handful of collectors' gardens in England and North America; it is hard to know why. During flowering time it is extraordinary, yet could be easily missed by casual observation. One good look will be enough to convince you that you have at once both a novelty and an exceedingly growable plant. While the blooms are unique, the noteworthy qualities of form and texture also have a lasting effect on both gardener and garden.

Geranium sinense creates a robust, mounded structure that associates easily and intimately with other perennials. It reaches at least 60 cm in height and width. Its lightly marbled, deeply cut leaves add embellishment at UBCBG, where they are featured in the foreground of a mixed border with an overhanging *Hedychium yunnanense*. The lightly shaded location satisfies this geranium's dual needs for dappled shade and moist soil. During midsummer it quietly harmonizes with its companions, but a bit later its unique floral quality comes to the fore. The flowers are rare in the plant kingdom by virtue of being nearly black, or a deep maroon-red to be exact. Adding to the mystery, the petals are strongly reflexed, and this inversion of the usual arrangement, together with the prominent nectary and filament display, creates an intriguing sight. In fact, the flowers bear a marked similarity to those of *G. phaeum*, but are not classified in the same taxonomical section because of differences in seed dispersal mechanisms.

Perhaps what fascinates us most about this plant is its unique juxtaposition of a familiar form with its strange flowers, together in an easily grown perennial. Because of its freely accessed nectar, this plant even attracts wasps rather than the common garden-variety bees, though I admit this will not appeal to some. This perennial is for anyone who loves to grow the astounding variety of taxa within this genus and wants to become acquainted with another easy, unusual one. Not only will you want to include this atypical Asian geranium in your collection, you may also elevate it to a prominent garden position.

HARDINESS: Zones 6b–9.

CULTIVATION: Due to its quiet nature and loose, lax habit, this species works best in plant associations instead of standing alone. Suggested companions include plants of erect habit, so this perennial may knit through them. Consistently moist soil in half-day sun also suits it best.

PROPAGATION: Geraniums are mostly quick to grow from seed, so there is really no need to bother the parent plant by digging and dividing it. On well-established plants, the root system resists easy division so it cannot be done without causing excessive disturbance. Plants must be carefully observed as the bloom cycle is ending, so that ripening seed can be collected before it is suddenly flung into garden oblivion. In our garden, this plant is not often observed to create seedlings.

Geranium soboliferum Komarov
Geraniaceae   Plate 44

DISTRIBUTION: Russia, Manchuria (China), Korea, and Japan, in meadows throughout various elevations.

RARELY ENCOUNTERED in cultivation in North America, *Geranium soboliferum* has long been known and grown in gardens throughout Europe. Considering its wide-ranging distribution over much of East Asia, it is no surprise that it performs so well in cultivation. It is also fortified against severe cold, which further enhances its garden profile. In spite of its cosmopolitan nature, it is a rare plant in the wild, where it inhabits sunny mountainsides. In our personal experience, it has also adapted admirably to the less-than-ideal meadow conditions present in the E. H. Lohbrunner Alpine Garden at UBCBG. The striking flowers and foliage are noticeably different among the bewildering array of choices available within the genus.

This perennial performs cheerfully in most good garden sites, particularly a perennial bed, where it can get all the nutrition it needs. Alkaline or acid conditions do not seem to make a difference if they are not extreme. In good conditions with moisture at the root in cultivation, the plant can reach a height of up to 80 cm on erect stems, but about 60 cm is the norm. Within its pleasingly mounded form, the leaves are deeply divided and lobed, lending a light, almost ferny appearance.

Its flowers also highly recommend it for general cultivation. They are proportionately large and distinctly dark red and purple. The bicolouration intensifies close to the centre of each flower, due to the strong crimson petal venation where the petals fuse. While the selection in our garden possesses this colour trait, plants elsewhere may not be as deeply infused. Dark red filaments set off the view of each of the copiously produced flowers as they dance open-faced among the foliage. The softly hairy, nodding buds also catch the eye and assure us of an ongoing supply from mid to late summer. The plant performs well in most garden situations receiving moderate to full sun and is best placed in the foreground, where it will captivate garden visitors.

Due to its evolution in grassy mountain meadows, *Geranium soboliferum* is a perennial competitor. Even so, it is well behaved and hasn't seeded about in cultivation at UBCBG. The species name *soboliferum* refers to the plant's formation of short, determinate runners that are its main means of slow increase. So far I can safely report that, though soboliferous, it has not become pestiferous. Because of its unquestionable hardiness, growability, and obvious beauty, this appealing *Geranium* species should prove popular in North American gardens.

HARDINESS: Zones 4b–9.

CULTIVATION: Like many meadow perennials, *Geranium soboliferum* is also easy to please, if its needs for good soil and a garden open to the sun and sky are met. While it may spread in cultivation (observation here shows that this is unlikely), more probably it will remain a distinct, clump-forming plant.

PROPAGATION: Do not count on the soboliferous qualities of this plant to make more plants. Once the plant has gained enough girth, it can be dug and divided with a pair of secateurs, during spring.

Hedychium forrestii Diels var. *forrestii*
Zingiberaceae   Plate 45

DISTRIBUTION: Southwest China, Laos, Myanmar (Burma), and Thailand.

THIS MOSTLY subtropical perennial has its center of diversity in Southeast Asia, at around latitude 20° north, so it is amazing that it performs well in the Pacific Northwest, more than twice as close to the North Pole. Essentially a plant of rugged lowlands up to about 900 m, it has remarkable hardiness. *Hedychium forrestii* was named to commemorate George Forrest, the prolific and determined Scottish plant collector. An astounding 90 years after he introduced this ginger to the West, it is still lingering on the frontier of our cultivation experience. In recent years, good new collections of the plant have been made, taking advantage of provenance (collection based on place of origin) and aiming to improve qualities such as vigour and hardiness. These new introductions have made available better selections for North American garden use.

This *Hedychium* species has been cultivated at UBCBG for more than 10 years, giving us insight into its robust, settled behaviour. It is planted in a woodland, so it has winter protection from occasional winter temperature extremes. *Hedychium forestii* comes mainly from southwest China, meaning a summer-rainfall climate, in contrast to the Pacific Northwest. In the garden, it is provided with a humic, yet draining, southwest-sloping soil that suits it perfectly. As a result, winter mulch, at first expected, is not needed.

Hedychium forrestii establishes its bold presence only after spring is over at UBCBG. It begins pushing up its remarkable 1.25-m stems rather suddenly, when the warmth of summer has really taken hold. Much later still, in early August, long terminal flower spikes expand. Two or three pure white flowers are borne in turn on the ends of tubular bracts. Exserted coral stamens complete the picture, and the flowers are slightly fragrant. From late summer to early fall, several seeds ripen within a capsule, which then splits open to reveal individual bright red arils. Later in fall, the leaves turn soft yellow, before the stems collapse at the end of the growing season.

Considering the trend toward growing subtropical plants, *Hedychium forrestii* is assured a bright future in North American gardens. It stands on its own, however, as a fine addition to the ranks of plants that were formerly considered to be tender.

HARDINESS: Zones 7a–9.

CULTIVATION: Superbly suited for use in woodland and other sheltered gardens and borders where there is moderate summer sun and regular moisture. Winter mulching is only necessary in marginal-climate gardens.

PROPAGATION: This perennial is quite easy to work with. Rhizomes are thick, reminiscent of store-bought ginger root, and grow just under the surface. In late spring, cut off a division with a sharp shovel and place it in a new home at the same level; it will quickly develop strong roots. Throw a dash of 6–8–6 fertilizer into the planting hole and stir, for good measure, as it will get the plant off to a good start. Mulch is unnecessary for our plants, which are grown in humusy soil in a sheltered woodland situation.

Hedychium yunnanense Gagnepain
Zingiberaceae   Plate 46

DISTRIBUTION: China, in Guangxi and Yunnan; India, Bhutan, and Indochina, including Vietnam, Cambodia, Thailand, and Laos.

THERE ARE about 50 species of *Hedychium*, and roughly two-thirds of these are native to Asia. In Victorian times, they were treated almost exclusively as tender perennials, reserved for glasshouse culture. Since that time, only a few of these were thought to be hardy outdoors in most of temperate North America. At present, however, this attitude has been overturned by successful trials of several species in gardens from California to southwest British Columbia.

Of the numerous *Hedychium* species growing successfully at UBCBG, my favourite is *H. yunnanense*, for three reasons. First is the plant's size. It is much shorter than most of the others, to about 1 m in our garden, which makes it easier to integrate into a wide variety of garden situations, from borders to balcony container culture. The natural tendency of *Hedychium* is to dominate its setting by virtue of its sheer size and volume, which is also true in cultivated landscapes. Second are the plant's luscious leaves, which are glabrous, broad, and blue-green, creating a tropical look. This "something extra" raises their glamour appeal, to be sure. Third, I am a fan of white flowers, and this perennial's display is as mouth-watering as vanilla ice cream on a hot day.

Similar to other *Hedychium* species, the growth cycle of *H. yunnanense* begins only after spring decidedly warms. No surprise considering its origins, except that at first planting, nothing seems to happen until a vague uneasiness has set in. Suddenly it awakens, and within a few warm weeks, leaves unfold, revealing its jungle charms. This species flowers earlier than many others, fully expanding its rounded, terminal inflorescences in late July. The pure white flowers are a sensual treat, since they have such magnificent fragrance—like gardenia—and the fully open flowers are exotically alluring. Individually, the spreading petals give an impression of a white bird in flight, but massed together, the open flowers look like exploding fireworks. Long exserted filaments recurve away from the petals, adding an incendiary touch. The illusion of the tropics is complete.

HARDINESS: Zones 7a–9.

CULTIVATION: Forests are the plant's preferred habitat, as long as they are reasonably temperate with draining soils. Mulch is only necessary in marginal climates, or right before unusual temperature extremes are expected to occur. Otherwise, these plants are best left alone, as they take time to settle after disturbance. In warmest climates, afternoon sun should be avoided, and moisture should be regularly supplied to the roots.

PROPAGATION: This perennial is yet another that is easy to work with. In late spring, cut off a division with a sharp shovel and place it in a new position at the same level; it will quickly develop strong roots. Mulch is unnecessary for our plants, which are grown in humus soil in a sheltered woodland situation.

Impatiens omeiana Hooker f.
Mount Omei balsam
Balsaminaceae   Plates 47–48

DISTRIBUTION: China, on Emei Shan, in Sichuan.

THIS HARDY plant belongs in a deservedly popular family containing some of the loveliest (and liveliest) perennials and annuals in cultivation. Until recently, though, it had proved challenging to track down attractive, easily growable, perennial species. But from among the lush undergrowth of Emei Shan (Mount Omei), one such plant was found. This distinguished species name celebrates a particularly hot spot in the plant diversity of China, an incredibly rich locale of botanical delights.

Modest in stature (to approximately 40 cm), *Impatiens omeiana* has more than enough presence to be highly prized in its cultivated woodland setting. In our garden, it spreads reasonably and is already active by late winter, pushing dusky red stems upward. By the middle of March, dark green, narrow, serrated, silver-patterned leaves serve notice that something special has arrived. These expand radially from many stems and cause comments from all who pass by. Variegation is an often sought-after woodland perennial feature (think of abundant white- and yellow-variegated hostas, legions of striped arisaemas and mottled epimediums, for example). By late summer and throughout a long fall, peculiar butter yellow flowers emerge, crouched atop the mini palm trees of foliage. This species works overtime to produce flowers even after the first gentle frosts, as long as days remain warm.

This plant's burgeoning popularity is based on its ease of culture and its charismatic combination of attractive foliage and exotic flowers in the midst of the dappled shade of a woodland garden setting. This perennial is just now appearing on the general horticulture scene and will undoubtedly have a solid future in North American gardens.

HARDINESS: Zones 6–10.

CULTIVATION: Thriving in leaf mould soil, the plant will spread slowly but surely via rhizomes to form a colony anywhere from 30 cm to 1 m in a couple of years. The plant's other underrated feature is its durability. It carries on undisturbed through the mainly dry Pacific Northwest summer (including very little irrigation) for a month or more, before flagging, and it moves faster in consistently moist soil. Unlike some of the genus's annual species, it has so far not become a pest by spreading its seeds by means of its exploding pods. We shall see how it behaves in warmer gardens.

PROPAGATION: As it grows best in the loosest and leafiest soils, it is easy to divide and move around the garden. Stems and rhizomes are brittle and need careful handling. Divide it in spring, a month after it appears aboveground.

Incarvillea compacta Maximowicz
Bignoniaceae   Plate 49

DISTRIBUTION: Southern Xizang (Tibet) and westcentral China, on slopes, in thickets and in grasslands, at 2600 to 4100 m.

JUST AS the indoor gloxinias were so thrilling to the Victorians, we are excited by the new "hardy gloxinias" recently arriving from Asia. Although the genus *Incarvillea* is compared with gloxinia due to a superficial similarity of their flowers, that is where comparisons end. *Incarvillea compacta* is proving to be hardy throughout much of North America. It is so named because its flowers are borne on a short raceme. Because of the species's nearly alpine origins, selections will probably prove to be even hardier in cultivation than has been described thus far, especially if plants are provided with some snow cover.

Incarvillea compacta arises from a thick taproot and forms a rosette of pinnate leaves during late spring. The leaves are wrinkled, and unfold as if they had been stuffed underground last fall. They are pinnately compound, with terminal ovate leaflets, a distinguishing character. Once soil has thoroughly warmed in late spring, plants produce dense 20- to 30-cm long racemes of sumptuous flowers, in shades from deep pink to reddish purple, depending on which clone is grown. Most plants now in cultivation have been gathered as seed from various plant expeditions, so there may be wide variation in available flower colour. The UBCBG accessions are a delightful shade of pink outside, and a lamp of bright golden yellow inside. The flowers may be expected to last for several weeks during late spring. On completion of bloom, seeds develop inside large pointed capsules, adding a lingering air of mystery.

Although coming from high elevations, *Incarvillea compacta* behaves well in cultivation in the northern maritime conditions of Vancouver. Like most others of its kin, this Asian perennial is a tough and beautiful example from a genus with a solid future in cultivation. Additional species are being steadily introduced, so we will no doubt see more of these sumptuous Asian ornamentals in gardens around North America.

HARDINESS: Zones 4–9.

CULTIVATION: Thus far, *Incarvillea compacta* has been planted at UBCBG in two sites, one in full sun, the other in half shade; both sites contain well-drained organic garden soil. In both situations, the perennial has proven itself adaptable for border use. The plant also appears to experience little trouble with pests, although during mild wet winters (where leaves often remain evergreen for an extended period), there has been some leaf and flower nibbling by slugs. The fleshy roots are rather brittle, so extra care must be taken when transplanting. In marginal climates, mulch and deep planting are beneficial.

PROPAGATION: This species and most others are taprooted, thus not allowing simple division. The practical propagation method is via seed, which fortunately is a rather rapid process. Expect blooming-sized plants within 3 years after planting seed.

Lilium taliense Franchet
Liliaceae   Plates 50–51

DISTRIBUTION: China, in Guizhou, Sichuan, Xizang (Tibet), and Yunnan, in open forests and on grassy slopes between 2500 and 3600 m.

LILIES ARE among many gardeners' favourite herbaceous perennials, for their unabashed beauty, and *Lilium taliense* is high on my personal lily list. Patrick Synge, in his book *Lilies* (1980) notes that this species was first collected in the Chinese province of Yunnan sometime in the 1860s, yet did not arrive in English gardens until 1935. Nearly 70 years later, I still wonder why this gorgeous, easily grown perennial has not yet caught on outside of botanical garden collections and a few private gardens.

A vigorous plant, *Lilium taliense* seems the antithesis of numerous lily hybrids grown in gardens today. In our garden, it has thrived in cultivation for almost two decades. Enjoying a forest-edge setting and healthy organic soil, it has responded by producing pure white and lemon yellow–blushed flowers on sturdy 1.4-m stems. In Yunnan, it is said to grow to an incredible 2.4 m. As many as 12 flowers per bulb, in the Turk's cap, or recurved, pattern, are produced in July, displaying a dark stippling of purple spots and flecks. It is a stunning sight to come upon this elegantly coloured, softly fragrant perennial in dappled shade, where visitors to UBCBG are yearly drawn to it. In essence, this lily is a restrained species that innately appeals to us while gaining wide appreciation as a consistent performer. Put another way, it is the epitome of gentle colour adornment in a peaceful setting.

Lilium taliense also performs well with other perennials and shrubs, such as dwarf rhododendrons, various clumping grasses, euphorbias, and large-leaf hostas. In such relaxed yet sophisticated mixed company, this Asian perennial will reward stewardship with many trouble-free years of robust growth and have your gardening friends asking you for surplus bulbs.

HARDINESS: Zones 5–9.

CULTIVATION: Because this plant primarily is a woodland lily, it should be placed where it can receive a half day of sunshine to do its best. Woodland-edge conditions are suitable, as long as the bed contains plenty of nourishing humus. The plant prefers regular summer moisture, but its roots cannot tolerate standing water: soil must be free-draining, but need not be specifically acid or alkaline.

PROPAGATION: Seed of this lily is produced in reasonable amounts, but the waiting time will test your patience. Instead, dig up a well-established clump and remove the young bulbs for planting in a new site that has been prepared with bonemeal and a dash of 6–8–6 fertilizer. If you are keen to obtain a mass planting of this lily for a fantastic display, you will need to use bulb scaling as a method of increase.

Linum hypericifolium Salisbury ex Steudel
Linaceae   Plate 52

DISTRIBUTION: Caucasus region, including eastern Turkey.

THE SCIENTIFIC NAME of *Linum hypericifolium* (hypericum-leaf *Linum*) does not do justice to this fabulous perennial. It is one of the highly ornamental flaxes that are new in North American cultivation, and it has certainly caught my interest. I started growing it at UBCBG in 2001 on a sunny slope in the E. H. Lohbrunner

Alpine Garden, which befits its origin on the steppes and high meadows of the Caucasus. In other words, it prefers a somewhat impoverished position in the garden to grow well, especially in the temperate maritime climate of the Pacific Northwest. I have heard that it is currently being grown successfully even in northern Norway.

So far this new flax has responded vigorously. It has developed along the lines of *Linum perenne*, by forming a robust clump in direct contrast to its spartan surroundings. During the warmth of late spring, it relishes the warm, stony soil and quickly develops erect and somewhat leafy stems to about 40 cm. Not long after, it bears 5-cm, bright rosy pink flowers at the stem ends. The colour of the flowers is so unexpectedly rich for what I have come to expect from a *Linum* that I occasionally find myself on bended knee admiring it. This plant has great potential for mingling with other bright or silver, drought-tolerant prairie-steppe perennials that also bloom in early summer. It will surely prove even more hardy than *L. perenne*, itself a handsome, long-blooming, and established standard in drier gardens. The many different species of *Linum* are well suited for growing in water-thrifty gardens throughout North America. This is also true of *L. hypericifolium*, which prospers in our garden.

HARDINESS: Zones 5b–9.

CULTIVATION: As much sun as possible, except in the hottest climates, average to mineral soil, and only moderate summer water are essential needs of this perennial, to ensure success.

PROPAGATION: Reasonable amounts of seed are found within capsules, and these will provide established progeny within 2 years. Fibrous root clumps eventually become tough with age, but younger ones are easily divided with a sharp knife in mid spring.

Nipponanthemum nipponicum (Franchet ex Maximowicz) Kitamura
Montauk daisy, Nippon daisy
syn. *Chrysanthemum nipponicum* (Franchet ex Maximowicz) Sprenger
Asteraceae   Plate 53

DISTRIBUTION: Japan, on central-eastern Honshu Island, near the ocean.

WELCOME TO the tongue-twisting world of botanical Latin. This subshrubby perennial chrysanthemum is another in the gardener's palette of valuable Asian daisies. Its Japanese name, *hama giku* (beach chrysanthemum), at least provides clues to its cultural requirements. I first came across the species years ago in the seedlist of Sendai Botanical Garden, Japan. It was described as a woody-based daisy, so I thought I had uncovered something different from the border chrysanthemums I already knew. Like other gardeners, I experiment constantly and try plants based simply on instincts, so I obtained and established this plant in the UBCBG David C. Lam Asian Garden, a mostly woodland garden. Knowing daisies crave plenty of light (and this one wants well-drained soil), I planted it in a partly sunny bed containing sandy loam. I was only half right, and the plant told me so. As it grew, it slowly bent over almost flat on the ground, reaching for those half-available rays. As much as it appreciated the dry soil conditions, it was in obvious need of more sun. Later, I planted it in the E. H. Lohbrunner Alpine Garden. In this exposed position, complete with its well-drained, lean soil, it has finally come into its own. In just 3 years, it has bulked up to substantial shrub-like proportions, and is adorned with glorious, pure white flowers very late in the season.

In spite of its late-flowering habit, this perennial creates an inviting picture from the season's start to finish. On erect 1-m stems, it creates a healthy image, clothed with entire, fleshy, glossy, serrated leaves. It looks like a Shasta daisy on steroids, because of its robust shrubby character. In our cool summer climate, it finally comes into its own during October. If our autumn is a sunny one, it will display copious blooms until late November. After bloom ceases, the plant behaves like a semievergreen shrub over an average winter in this climate until spring sees it bud out again. This perennial is one of the rare ones with the ability to perform as well out of bloom as in. It is a daisy of distinction, having more than enough fine qualities to earn pride of place in warmer North American gardens.

HARDINESS: Zones 7a–10.

CULTIVATION: Cultivation is simple, although somewhat different from most other daisies. Cutting the woody stems back by half in late winter or early spring encourages buds to break, creating a denser plant and improving fall flowering. Plants can also be pinched in early summer to increase flower production, an old chrysanthemum-fancier's trick. Where winters are more severe, the plant will die back to the ground each year. It is now being grown in some gardens on Long Island in New York, and I am certain it must feel right at home there with those hot, humid summers, just like in Japan.

PROPAGATION: Careful division can be performed in spring, and semiripe cuttings strike easily in mid to late summer.

Paris fargesii var. *petiolata* (Baker ex C. H. Wright) F. T. Wang & T. Tang
Trilliaceae   Plates 54–55

DISTRIBUTION: China, in Guangxi, Guizhou, Jiangxi, and Sichuan, at 1200 to 2000 m, in shady forests.

BECAUSE we regularly admit many unusual and striking wonders of the temperate plant world to our garden stewardship at UBCBG, we can easily become a bit jaded in our horticultural tastes. Among the potential winners we receive for trial, we are universally ecstatic about this fabulous *Paris* species from China. Taxonomically classified in Trilliaceae, this genus bears superficial floral resemblances to its somewhat showier cousins, the trilliums. The genus also shares a similar environment—shady, moist woodlands. Here in the garden, this species revels in its position in early morning sun, followed by shade for the rest of the day. This variety, *petiolata,* is so named because of its especially long petioles (much longer than the species), extending up to 10 cm from the main stem. From these, 6 or 7 leaves are set in a whorl about 30 cm above the ground. This umbrella-like appearance is the plant's foremost feature. Both leaves and sepals have downward-pointing tips that hint at the plant's evolution as a rain-forest species. Yearly developing into a more impressive specimen, our plant now reaches a spectacular 1.5 m in height.

This perennial's flowers are as delightful as they are unique. They are borne singly atop each sturdy stem, displaying 5 or 6 large, shiny, green sepals. Above these are tiny, linear, yellow petals, which originate from below a prominent dusky ovary. By July, the flowers have matured, but because their development is spread over such a long period in spring, they garner a lot of attention. Planted near the front of the border, these examples provoke a steady stream of positive comments from passersby. This Asian *Paris* is a definite must-have for the garden.

HARDINESS: Zones 6–9.

CULTIVATION: This woodlander requires humus and draining, moist soil. The robust nature of these plants is another strong point. They settle down quickly in cultivation and develop well without a hitch. Within 3 years, our specimen has spread by rhizomes to form a strong clump, bearing several flowering stems. Annual mulching with well-rotted compost encourages better growth. The plant has performed admirably in our region's somewhat acid loam, although it prefers a neutral balance to do its best.

PROPAGATION: This perennial is mainly clump-forming, with a tendency to develop very short rhizomes. The rootstock eventually becomes tough, so it is best to divide plants carefully while they are relatively young; this can be done after bloom, in early summer. As they resent disturbance, this will allow parent plants to recover more quickly. In good garden conditions, new plants may be expected to reach bloom size in 4 or 5 years.

Persicaria tenuicaulis (Bisset & S. Moore) P. Trehane
Polygonaceae   Plate 56

DISTRIBUTION: Japan, on Honshu, Shikoku, and Kyushu Islands.

NAMED FOR its slender stems, *Persicaria tenuicaulis* begins to leaf out very early in the spring. Its ground-covering habit leads it to form a distinctive dark green carpet among the leaf litter. Due to its prostrate habit, it reaches only a modest 20 cm in height. By mid-spring, bright white, foaming-toothbrush, flowering spikes appear in profusion in scale with the plant. Stamens exsert from the corollas and add a fuzzy appeal when viewed closely. The central, dark red, flowering scapes offer a modest pink blush. Flowering also occurs early, so the plant should be placed where it can be appreciated. Later in the spring and early summer, the petioles of the small leaves elongate to about 20 cm. By this time, the plants have formed leafy clumps that have begun to merge with each other. The rhizomatous habit is gentler than that of its more aggressive cousin, the widely used *P. affinis*.

This plant belongs to a desirable and gardenworthy group of polygonaceous perennials already popular throughout North America, but *Persicaria tenuicaulis* has thus far been overlooked. The plant is a suitable, charming inclusion for dappled shade gardens, patios, and other small but horticulturally useful spaces.

HARDINESS: Zones 6–9.

CULTIVATION: In our garden, only a modest bit of slug damage has been observed on the leaves, but not nearly enough to detract from the plant's overall picture of health. Apparently, this perennial is no more attractive to the dreaded mollusk than most other perennials. No other problems with insects or disease have been noted, and our plants are now 10 years old.

PROPAGATION: The rhizomatous root system is not too tough, so it is easy to separate a piece from the parent. Divisions made in spring, when the soil is moist, result in thriving new colonies of this polygonum.

Persicaria vacciniifolia (Wallich ex Meisner) L. P. Ronse Decraene
Polygonaceae   Plate 57

DISTRIBUTION: The Himalayas, from Pakistan in the west to Bhutan in the east, at 3000 to 4200 m, in thickets and rock crevices.

A DIMINUTIVE subalpine gem, this plant has a host of attractive qualities. It is one of those plants that people seem to know about, but not many (outside of rock gardening circles) seem to grow. This woody-based, ground-covering, polygonaceous performer is a great choice for gardens requiring little maintenance.

In contrast to some commonly available bigger species of *Persicaria* that tend to spread into their neighbours, *P. vacciniifolia* is petite and remains contained. In the relatively dry E. H. Lohbrunner Alpine Garden, it spreads no farther than 1 m or so, with age. Slowly creeping, it carpets bare ground while smothering most small weeds. It comes from mountainsides in the Himalayas, where it grows sturdily, all in a short growing season. In cultivation at UBCBG, it is grown in gritty mountain-type soil and has never lost its well-behaved habit, even after more than 25 years. Our specimen overhangs a large boulder and looks completely at home in its surrogate environment. This perennial has one of the distinctive characters of choice plants: the ability to adapt to a cultivation setting different from its native habitat.

By late summer the decumbent habit takes on a new dimension as the flowers emerge on multitudinous spikes. They are bright, soft pink on 30-cm stems, both floriferous and subtly welcome as they usher in a new autumn season. The stems are well clothed in small linear leaves. As its species name suggests, the leaves of this plant are somewhat reminiscent of *Vaccinium* leaves, even to the extent that they attractively colour in rich tones of orange and red before falling.

HARDINESS: Zones 6b–9.

CULTIVATION: Although this plant prefers moist soil, it requires only a modest amount of moisture in the growing season to look well; its leafy, wandering stems are an effective means of limiting soil moisture loss.

PROPAGATION: The easiest method is simply accomplished by digging a few stems from where they have rooted into soil and detaching these from the parent plant. Another method is by softwood cuttings, taken in early summer. By either means, you may give a piece of this first-class herbaceous character to a friend.

Phlomis cashmeriana Royle
Kashmir sage
Lamiaceae   Plate 58

DISTRIBUTION: Kashmir (India) and the western Himalayas, in a variety of open situations.

EVERY GARDENER I know loves *Phlomis*. There is something about this genus's soft, fuzzy look that gives it universal appeal. *Phlomis cashmeriana* is almost unknown in North American gardens, perhaps due in part to the long-standing reliance on *P. fruticosa* and *P. russelliana*, both large, yellow-flowered plants from the Mediterranean. For something quite different yet equally easy to grow, *P. cashmeriana* is a fine choice. Technically classified as a subshrub, it performs as a woody-based herbaceous perennial at UBCBG. It hails from the drier, western Himalayas and Kashmir, as its species name suggests, so it is best suited to sunny and well-drained garden sites.

Phlomis cashmeriana has plenty of appeal as both a focal point in a dry border (it performs best here in the E. H. Lohbrunner Alpine Garden) and as a mingler and softener among flashier drought-tolerant plants. Like other *Phlomis*, it is a natural for the xeriscape garden, and is already being recommended by Denver Botanic Garden. Emerging suddenly into the spring sun with intensely silver and hairy new leaves, it quickly establishes its handsome architecture. By summer it puts on about 60 cm of growth, although it can reach 80 cm or more in favourable gardens. In this exuberant, grey-green mass are many sturdy, white-felted stems that bear, in the peak of summer, several verticillasters (whorled clusters) of exotic lilac-purple flowers.

This plant, with its subtle colour combinations of silver mixed with pink and purple, calms the senses and invites our touch, while standing uniquely apart from other dazzlingly bright summer flowers. After a few weeks, flowers finish blooming and stems may be cut back, leaving a superb foliage plant, untouched by insect or disease problems, for the rest of the growing season. Stems also may be left on, providing textural interest into the fall and winter seasons. This Asian perennial is both showy and not easily bothered by extremes of climate. Its appeal centers around its strong form, combined with a touch of the exotic.

HARDINIESS: Zones 5b–10.

CULTIVATION: Our plants are located in well-drained beds receiving full sun. In winter, the perennial usually becomes a nondescript pile of dead foliage that serves to protect the crown. During the growing season, its needs only a modest amount of water. This subshrub would be a good foundation plant in a xeriscape garden.

PROPAGATION: Seed is an option, but cuttings taken from early summer onward are a quicker option.

Phytolacca clavigera W. W. Smith
Chinese pokeweed
Phytolaccaceae   Plates 59–60

DISTRIBUTION: Southwest China, in Yunnan, in moist sites, at low to middle elevations.

CHINESE POKEWEED is a rare temperate member of a largely subtropical and tropical Northern Hemisphere genus and family, with some members in the Southern Hemisphere. The genus *Phytoloacca* comes from *phyto* meaning "plant" and *lacca* meaning "lacquer," and the epithet *clavigera* means "club-bearing," a reference to the flowering racemes. Although this plant may appear similar to the widely known *P. americana* (aka pokeweed), the Chinese species is more refined than its New World counterpart. While keeping to a more compact 1.5 m, its dense flower clusters and berry display are strikingly ornamental. There is also a Chinese counterpart, *P. acinosa*, but it does not offer the ravishing qualities of *P. clavigera*, which carries much longer flowering racemes. If these three plants were grown side by side, it would be immediately apparent that *P. clavigera* is the outstanding plant. It is a versatile, handsome, and extraordinary display perennial that often has visitors to UBCBG enquiring, "What is that plant?"

I have valued the qualities of *Phytolacca clavigera* for more than 10 years in the perennial border at UBCBG. Technically classified as a subshrub, it behaves as a herbaceous perennial here, its husky, deep purple stem deriving from a solid, woody rootstock. Although by late spring it quickly develops a shapely form, from midsummer onward this plant reaches its most appealing stage. The sturdy open framework is clothed in 15-cm, soft-looking, ovate leaves. One quickly notices its spectacular flowering racemes, which are densely formed, to 30 cm (much longer than in *P. americana*), in a luscious pink, with green berries soon following.

By late August, the fruits have ripened to shimmering black clusters that resemble small blackberries, and these are combined with bright pink pedicels. I have learned the hard way that the fruits have the same staining ability as ink. In fact, they are used as a dye in China. *Phytolacca* fruits are relished by birds in the southeastern United States, a major source of the plant's spread there. Although birds may ingest them harmlessly, the berries and roots are poisonous to people. Even small amounts can cause major problems, especially for young children. In climates warmer than the Pacific Northwest, there will probably be some seedlings noticed around the garden, by the same dispersal method. The young leaves of both *P. clavigera* and *P. americana* are palatable after cooking (which destroys the toxic alkaloid, phytolaccin), and the latter has been used as a pot-herb for years. Both are also known to have a certain quickening effect on the body. In our typical mild fall, the leaves respond to cooler nights and shorter days by taking on a warm yellow glow.

Phytolacca clavigera is conspicuous for its handsome habit, stunning flowers, and striking fruit display.

HARDINESS: Zones 6–9.

CULTIVATION: This plant grows robustly in any good garden soil. In a sunny, moisture-retentive border habitat, it grows rapidly and luxuriously.

PROPAGATION: The plant's rootstock resembles the bole of a small tropical tree more than a herbaceous perennial. The most direct means of propagation is by seed, which is produced copiously. Seed sown when ripe results in new plants that could be over-wintered under cover and planted in permanent positions the following spring.

Potentilla cuneata (Wallich) J. G. C. Lehmann
Rosaceae   Plate 61

DISTRIBUTION: China, in Sichuan, Xizang (Tibet), and Yunnan Provinces, and in Bhutan, Nepal, and India, in Sikkim, in forest edges and alpine meadows, at 2700 to 3900 m.

AMONG THE indomitable dwellers of the higher reaches, the genus *Potentilla* truly stands out. It is a vast group, and *P. cuneata* is a fine example. Technically a sub-shrub, this groundcover behaves in cultivation like a herbaceous perennial. The specific epithet refers to its leaf bases, which taper down (wedge-like) to the peti-oles. In its native habitat, it grows on exposed hills and meadows, slowly creeping during the short growing season. What makes this perennial so eminently suit-able in cultivation is its ability to adapt to and thrive under markedly different conditions. A prime example derives from a seed collection taken in Nepal at 3870 m. Coming from what can be an unforgiving climate for most of the year, our plant has been thriving in cultivation at UBCBG for 30 years. Locally, it expe-riences a mostly snowless climate at close to sea level. So this little fellow has proven itself a marvel of adaptability. Its value is further evident in its multisea-son good looks, including an extensive bloom of golden yellow flowers.

The short rhizomes of *Potentilla cuneata* slowly form a close-knit carpet in gritty soil, and after many years the plant has reached a diameter of about 2 m. Meanwhile, from midsummer the extended flower show begins. The bright blos-soms are short-stemmed and produced in such profusion as to turn the plant into a reflection of the sun. Although the main flowering season of these 2.5-cm sin-gle "roses" is about 6 weeks, intermittent bloom lasts until the cooler nights of fall. The perennial has charming trifoliate, silky, incised leaves that exhibit a warm-toned, colourful display before they drop.

HARDINESS: Zones 5–9.

CULTIVATION: This plant needs only three things to make it happy: a sunny site, mineral soil, and steady moisture at the root. Okay, cheat a little and add a dash of 6–8–6 fertilizer during spring. You may anticipate years of reward from this tough and beautiful perennial ground cover.

PROPAGATION: The simplest method of increase is to plunge a spade into the col-ony near an edge and remove a well-rooted section to its new home. Seed is pro-duced, but it is minute and is best gathered (if it all) to be used for seed exchanges with friends.

Ranzania japonica T. Itō
Berberidaceae   Plate 62

DISTRIBUTION: Japan, in Hokkaido and westcentral and northwestern Honshu, in mountain forests.

THIS UNIQUE Asian woodland jewel was named in honour of the Japanese botanist Ranzan Ono when it was described just over 100 years ago. Since becoming scientifically recognized, it has been moved from Podophyllaceae into its current classification in Berberidaceae. From a gardener's viewpoint, it is quite distinct in both flower and foliage.

The best time to observe *Ranzania japonica* is during the warmth of spring, when it produces its new growth. As its leaves unfurl and it assumes its 45-cm height, nodding, pale lavender-pink flowers appear from long, narrow pedicels. They are a visual treat and best appreciated from down low where they shyly open. Referring us again to the family Berberidaceae, the flowers are reminiscent of *Jeffersonia dubia*, another choice perennial of the northeastern Asian woods. During late spring, the ternate leaves of *Ranzania* enlarge into cordate-acuminate parasols with a pleasing lime-green colouration. On their own, they create a stir, as they resemble no other plant in the garden, except perhaps the native vanilla leaf, *Achlys triphylla*, itself another herbaceous berberidad. For this reason, *R. japonica* is worthy of a prominent site in the garden, because it will continue to intrigue viewers even after its enchanting floral act concludes. By summer, a pearl-white berry is produced under the leaves, which lasts a short time before opening to disperse its seed.

Ranzania japonica comes from mountainous northwestern Honshu, where frequent snows and substantial cold are common during winter. While the Pacific Northwest is no real testing ground of the hardiness of many Asian perennials, this one is tough enough to withstand at least a Zone 6 winter with modest protection. This plant has not become well known here yet, perhaps because it is so rare in the wild. But it is not difficult to grow and propagate. And the garden future of this true Asian aristocrat is assuredly bright.

HARDINESS: Zones 6–9.

CULTIVATION: This perennial likes a humus-rich bed and afternoon shade, because its natural habitat is the understory of mostly deciduous woods. Treat it like you would most trilliums and you will find the plant an easy-care perennial.

PROPAGATION: Seed can be gathered from down under the leaves, from the little lidded capsules, and it should be sown immediately. Division with a sharp knife can be performed on older plants. *Ranzania japonica* enlarges slowly and may take more than 5 years before a reasonable clump develops.

Rehmannia glutinosa Liboschitz ex Fischer & C. A. Meyer
Chinese foxglove
Scrophulariaceae   Plate 63

DISTRIBUTION: China, in Gansu, Hebei, Henan, Hubei, Jiangsu, Liaoning, Nei Mongol, Shaanxi, Shandong, and Shanxi, on hillsides and mountain slopes to 1100 m.

A SMALL yet striking perennial, *Rehmannia glutinosa* belongs to a family appreciated for its extraordinarily beautiful flowers. Its "big brother," *R. elata*, has trailblazed into North American gardens, flashing its large, deep pink flowers along 1.5-m stems. Since then, *R. glutinosa* has quietly come along and is creating a stir of its own.

It is compact, to about 30 cm, and gently pushes its rhizomes through well-drained organic soil, creating small colonies. By its second season, it is likely to show a half dozen 30-cm flowering stems, and by its third or fourth year, it may spread to 1 m or more in diameter. As such, it is best suited to garden sites where it may freely roam. Typically in the Pacific Northwest, every few years we receive deeper frosts, which curtails further expansion. Combined with soaking winter rains, a colony can be effectively reduced to just a couple of overwintering individuals, and then will begin the cycle again.

The leaf character of *Rehmannia glutinosa* includes veins that appear stamped on, and they angle upward at 45 degrees. The entire plant and especially the stem is covered in soft purple hairs, spreading into the throats of the unique flowers. The corolla is narrow, but the salverform, lobed petals flare brightly outward. Flower colour is variously described as brick red, red-brown, or even brownish purple-yellow (this is quite accurate in some individuals), providing them with instant eye appeal. They form in the leaf axils on short stems and on terminal racemes. Essentially, *R. glutinosa* is a free-flowering mid-spring bloomer with intermittent flowering until midsummer. Its distinctive colourful flower display and sticky, hairy habit entice garden visitors. Given freedom of movement in cultivation, it requires no special treatment to prosper. In Chinese herbal medicine, its root is used for various treatments.

HARDINESS: Zones 7b–9.

CULTIVATION: Healthy proportions of both grit and humus in the growing medium are essential to this plant's well-being. Currently, plants grown in various areas of our garden are placed where they can receive at least a half day of sun. One plant is sited among dwarf rhododendrons, whose compact root systems are not bothered by the wandering rhizomes of *Rehmannia glutinosa*. Although this species manages well on the West Coast, its origins suggest suitability for North American gardens with a summer rainfall regime and drier winters than our own.

PROPAGATION: Because the plant spreads by rhizomes, it is a simple matter to cut away a few roots and spread them out into humus-rich soil. Seed is usually plentiful (though tiny) and must be sown in early spring to ensure large and well-rooted plants by the following fall. For the impatient gardener, cuttings may be taken in early summer, but success is only just worth the effort. Seed propagation produces more progeny in a reasonable length of time.

Roscoea tibetica Batalin
Zingiberaceae　　　　　　　　　　　　　　　　　　　　　　　　Plate 64

DISTRIBUTION: China, in Sichuan, Xizang (Tibet), and Yunnan, and in Bhutan, India, and Myanmar (Burma), at 2400 to 3800 m; not considered alpine, but a part of the adjacent coniferous zone.

THIS HARDY perennial is a member of a lovely and intriguing family with orchid-like flowers and fleshy roots. Plants are short, sturdy, and tuberous-rooted, and live in a variety of habitats from moist meadows to dry grassy hills and forest clearings. They seem unfussy about their companions and can be easily missed in spring because they rise late, but by summer, they offer their unique, beautiful flower displays.

Although no more than about 20 cm tall, *Roscoea tibetica* demands to be noticed. The plants at UBCBG are situated near a path where they can be easily viewed. The flowers are enticing. Produced on terminal spikes, they resemble some orchids in their petal arrangement, including a posterior hood, two spreading laterals, and a protruding, pouting lip. The colour is a deep ravishing purple. Several flowers are produced, beginning in our garden in July, one at a time, and continuing until near summer's end. Later, a seed capsule is formed (again reminiscent of Orchidaceae), which eventually dehisces to reveal many dark seeds within. This plant forms leafy rosettes and increases slowly by rhizomes.

HARDINESS: Zones 6b–9.

CULTIVATION: *Roscoea tibetica* benefits from mulch to aid its root development and subsequent flowering. It appreciates partial sun and conditions that will not allow roots to dry out, which can quickly spell their demise. Once planted, they should not be disturbed.

PROPAGATION: *Roscoea tibetica* roots are fleshy and brittle, so they should be handled carefully. Clumps can be dug in spring (as growth buds show above the soil), or in fall in mild climates. Carefully tease apart individual plants and replant in humic soil. Colonies are usually long-lived. Seed may be collected and sufficient amounts will be found inside elongated capsules among the leaves. As *R. tibetica* originates from montane elevations, it may prefer a cold period to break dormancy. Seed may be stored in the refrigerator until the following spring before sowing.

Roscoea wardii Cowley
Zingiberaceae

DISTRIBUTION: China, in Xizang (Tibet) and Yunnan, and in India and Myanmar (Burma), at 2400 to 3500 m; not considered alpine, but a part of adjacent coniferous zone.

As PART of the ginger family (Zingiberaceae), the genus *Roscoea* is a sophisticated group of perennials. These plants have survived longer than many other flowering plants, and many of the temperate and even subtropical individuals are tougher than most of us currently believe. Both *Roscoea tibetica* and *R. wardii* are newly introduced species that will soon enjoy popularity in warmer gardens throughout North America.

   Roscoea wardii is similar to *R. tibetica* but more substantial, reaching 30 cm tall. From its clasping leaves, clusters of soft, satin-pillow purple flowers are produced in early summer. They are also proportionately larger than those of *R. tibetica*, yet with the characteristic hooded appearance.

HARDINESS: Zones 6b–9.

CULTIVATION: *Roscoea wardii* benefits from mulch to aid its root development and subsequent flowering. Plants appreciate partial sun and conditions that will not allow roots to dry out, which can quickly spell their demise. Once planted, they should not be disturbed.

PROPAGATION: The fleshy, brittle roots should be handled carefully. Clumps can be dug in spring (as growth buds show above the soil), or in fall in mild climates. Carefully tease apart individual plants and replant in humic soil. Colonies are usually long-lived. Seed may be collected and sufficient amounts will be found inside elongated capsules among the leaves. As *Roscoea wardii* originates from montane elevations, it may prefer a cold period to break dormancy. Seed may be stored in the refrigerator until the following spring before sowing.

Salvia koyamae Makino
Japanese yellow sage
Lamiaceae   Plate 65

DISTRIBUTION: Japan, on Honshu Island, where it is localized and rare.

IN MANY North American gardens, we often see the same few sages grown year after year. *Salvia koyamae* (pronounced ko-yam-ee) is one of many untried Asian species, possessing some desirable traits that will provide gardeners with unique and interesting cultivation options. Known in the West by its common name Japanese yellow sage, there is much to know about it.

Not a typical sage, this genus needs shade to prosper. In fact, it will not grow well in a sunny environment. A cool inhabitant of the forest understory, in its congenial setting it performs as a happy perennial woodlander. At UBCBG it has so far proven adaptable enough to accept modest amounts of morning sun or dappled shade. Because of this adaptation to low light, its habit is atypical. It is not a plant of erect habit, aimed squarely toward the sky, but instead forms a leafy ground cover that over time can spread to 1 m or more in width. In height, it can reach up to 80 cm in rich organic soil, which it prefers. Our plant has attained a more modest 60 cm in average soil conditions. The thick stems of *Salvia koyamae* are more or less decumbent, or lax. But the terminal flowering spikes become vertical, allowing them to be shown to advantage.

As the stems grow, they root at the nodes, thus forming a colony. The leaves are classically sage-like, deltoid, quite large (to 15 cm long), and neatly serrated, lending ornamentation even when the plant is out of flower. The flowers are beautiful, pairing effectively with the foliage. Hued in pale yellow, they delicately punctuate their wooded backdrop. Beginning in late summer and lasting well into fall, the plant blooms for several weeks. At that time of the season, those who explore our shady garden pathways enjoy discovering this rare and intriguing Asian perennial.

HARDINESS: Zones 6–9.

CULTIVATION: This plant has demonstrated its adaptability to the mild but seasonally damp growing conditions in the Pacific Northwest (it prefers regular moisture), having performed well since 1996. Coming from Japan, it will assuredly be a natural for summer-humid regions of southcentral Canada and the eastern seaboard and southeastern region of the United States. At home in summer subtropical moisture, this plant is bound to exhibit optimum overall performance in similar regions.

PROPAGATION: This species does not produce an abundance of seed in the local conditions, but it roots at the surface as it spreads. With a garden fork, lifting its stems is easily accomplished, so they can be separated and replanted elsewhere.

Saxifraga stolonifera Curtis '**Cuscutiformis**'
Saxifragaceae   Plate 66

DISTRIBUTION: Southern and eastern China, from sea level to 4500 m.

THIS PERENNIAL is a very hardy and robust form of *Saxifraga stolonifera* (commonly known as "mother of thousands") with an odd name. Since it was described, it became linked in people's minds with the genus *Cuscuta*, a group of parasitic, strangling vines. The numerous, red, thread-like stolons resemble the brightly coloured stems of *Cuscuta*. However physically accurate, this explanation does no service to what is an attractive and eminently growable species. This plant also should not be confused with the similar (smaller and less hardy) *S. cuscutiformis*. Having gotten these tricky details out of the way, the comparisons end. This special form provides North American gardens with a handsome, free-flowering perennial that is simple to work with, providing you meet its basic requirements. It has been classified in section *Irregulares* of *Saxifraga* in *Flora of China* (Wu and Raven 2001), which includes most of eastern Asia's forest dwellers.

This perennial is one of the most attractive members of its woodland group. During the warming days of spring, small, overwintering rosettes expand to form dark green, sparsely hairy leaves on short, red stems. Pale silver veins mark the upper surface of multiplying, kidney-shaped leaves. Small, red stolons then elongate, and the plant creates moderate-sized colonies over the summer. Its delightful character can be appreciated whether in a cosy patio garden setting or en masse in a woodland context. Although it tolerates the summer drought at UBCBG, it prefers a humic, moisture-retentive soil. High soil fertility is essential for its well-being, providing the nutrition necessary for robust behaviour. White flowers are formed on branched panicles during late summer, rising to 30 cm and suspended on dark red stems, appearing to float like white butterflies amidst their shady surroundings. They positively entice viewers to closer inspection and should be considered among the finest of all perennial flowers, each one a simple work of art.

Due to its captivating qualities and ease of cultivation, this plant is extremely gardenworthy. Most gardeners have already seen *Saxifraga stolonifera* in cultivation and it has adapted well to houseplant culture. Outside in the shaded garden, effective pairings with a selection of deciduous rhododendrons, skimmias, or camellias would allow it freedom of movement while highlighting its singular beauty. *Saxifraga stolonifera* 'Cuscutiformis' is an exceptional perennial awaiting those who seek a beautiful ground cover to enhance moist, fertile shade gardens.

HARDINESS: Zones 5b–9.

CULTIVATION: Selections of this superior plant will probably perform better in summer-humid conditions common in eastern North America than in the summer-dry western North America. Among widely differing North American soils, this plant is adaptable to acid or alkaline conditions, making it all the more valuable for garden use.

PROPAGATION: The simplest way to produce more of this lovely perennial is, during late spring's growth spurt, to cut away rooted rosettes, including stolons, from the parent and move them to other places in the garden. As long as the new site is mostly shady and soil remains moist while the offsets are reestablishing, the plants will quickly settle down.

Senecio cannabifolius Lessing
Aleutian ragwort
syn. *Senecio palmatus* Pallas ex Ledebour
Asteraceae   Plate 67

DISTRIBUTION: Korea, eastern Russia, northern Japan, and east to the Aleutian Islands (United States), in low and higher mountains and wooded openings.

FLIGHTS OF FANCY with its species name well aside, this distinctive perennial has a robust habit. I am impressed with its unusual architecture, and as a plantsman, I am immediately attracted to it. During my last visit in Japan, I had the good fortune to see this species growing wild in the Japanese Alps on Honshu Island. Upon first observation, it carried an independent demeanor in its position in a bright, open wood, prominently crowned with corymbs of yellow flowers. I felt that it exuded something primitive and mysterious. Considering its distribution in other more remote regions, such as Sakhalin and the Aleutian Islands, it is perhaps appropriate that it conveys such an untamed air.

On its 2-m stems, scattered and widely spaced palmate leaves project horizontally and hang limpidly in the air. They are pointedly 3-lobed and prominently veined, giving a fleeting impression of slender human hands. The stems themselves are strongly erect and reddish tinted. Bright yellow flowers appear in midsummer, after which long-lasting, fluffy seedheads form and cling, prolonging interest well into crisp fall days. On the practical side, most *Senecio* species are similar in their cultural requirements to *Ligularia*, and our accession is happily sited in a moist position in a forest opening, duplicating its natural environment. This trouble-free, bone-hardy plant is well suited for the wild or bog garden.

HARDINESS: Zones 5–9.

CULTIVATION: Since it wanders liberally by rhizomes, give *Senecio cannabifolius* the room it needs. Plant it in a sunny, moist position on its own, and enjoy seeing it roam, much as it freely does in the wilds of northeast Asia.

PROPAGATION: Copious viable seed is produced, but it has not yet produced any seedlings in our garden. The rhizomes are more than enough to increase the stock. Rooted rhizomes can be dug from leafy soil most anytime from spring until fall and replanted where desired.

Strobilanthes attenuata (Nees) Nees
Acanthaceae   Plates 68–69

DISTRIBUTION: Nepal, the Himalayas, including north India, in foothills and montane slopes.

THIS SPECIES is a member of the 250-strong herbaceous family Acanthaceae, which includes the well-known genus *Acanthus* (bear's breeches). *Strobilanthes attenuata*, however, is nothing like that, at least on the surface. At UBCBG, this plant suddenly distinguishes itself in late spring from the native background of *Thuja plicata* (western red cedar), which is of a similar lime green folial colour. Because it rises erect on branched stems that are hairy and jointed as well as squarish in cross section, *Strobilanthes attenuata* is understandably compared with *Salvia*. It reaches about 1.5 m, full of relaxed, entire leaves, which are cordate and hang oppositely from the stems. They make a fine display, as they are handsomely corrugated and drip-tipped. The specific epithet brings attention to this adaptive attribute, its leaf-tip attenuation, a common characteristic of life in the monsoonal region of the Himalayas. In spite of its preference for moisture, it has performed quite well for a number of years in its summer-dry, half-sun position in the David C. Lam Asian Garden.

This perennial is especially attractive during its copious flower display. When most other perennials are already tired, it begins flowering in midsummer, producing a long and satisfying show. Bursting forth in axillary and terminal spikes, each flower is formed into a delicious, concord grape-purple, semicircular tube. Each is closed at the terminal end at first, later flaring open to reveal a lighter interior. Its splendid summertime combination of foliage and flower colour—lime green and rich purple—is cool and pleasant.

HARDINESS: Zones 5b–9.

CULTIVATION: *Strobilanthes attenuata* spreads gently around its garden bed (in soil of average fertility) by short, clumping rhizomes and by seed. Its water needs are modest, yet it will flag if left without water for too long. A position in well-drained yet moist soil will promote its enthusiasm for cultivation.

PROPAGATION: This plant is a stout and splendid clump former. I have found, however, that it is somewhat prone to adding seedlings to its family group at its feet, which are easily moved into new homes the following spring as desired. It is not at all a weedy plant, but requires placement in an accommodating garden. It looks best as a massed display.

Veronica ornata Monjuschko
Scrophulariaceae   Plate 70

DISTRIBUTION: Japan, on Honshu Island, in scrub and grassy places at lower elevations.

THIS BEAUTIFUL herbaceous perennial, coming from the shores of Japan, provides so much interest and garden value that it is bound to cause attention among even the most jaded gardeners. With the epithet *ornata*, you know it has to be good.

Veronica ornata grows to about 60 cm, bearing frosted stems and silver-green, opposite leaves. It is reminiscent of *V. spicata*, a European speedwell thoroughly represented horticulturally by various selections. However, *V. ornata* creates a bigger splash in gardens, with better leaves and richer flowers. Its native habitat is described in Ohwi's *Flora of Japan* (1965, 799), as "pine woods near seashores," so with this prescription for success in mind, three individuals were planted in the E. H. Lohbrunner Alpine Garden at UBCBG, under a *Pinus parviflora* (Japanese white pine). In this summer-dry, exposed site, it has flourished.

In spring it rises quickly, forming neat clumps of soft-focus, grey-green leaves. By mid August, 25-cm flower spikes burst forth, heralding the flowers, which are an exquisite twilight blue, densely held aloft on multiple firm stems. As the coming fall weather cools, this magnificent silver-green to sapphire-blue display continues well into the fall, even into December in some years. Among the veronicas, it has an unequalled sophistication. It is a winner for the late summer and fall garden.

HARDINESS: Zones 6b–9.

CULTIVATION: This floriferous perennial is easy to please. Although it has performed well at UBCBG in dry, mineral soil conditions, it probably would prosper in even greater measure in a sunny site containing good garden loam. In spite of its less-than-optimum conditions, it produces an abundance of flowering stems.

PROPAGATION: Placed in sunshine and good soil, this perennial soon reaches sufficient size for division. Since the root system is very fibrous, separating can be easily accomplished either in spring or early fall. Cuttings may be taken as well.

PART 2
Shrubs

Peter Wharton

Alangium platanifolium (Siebold & Zuccarini) Harms
Alangiaceae (Cornaceae)   Plates 72–73

Distribution: Japan and China, at 1300 to 3000 m.

This outstanding understory shrub can become a small, wide-spreading tree of considerable beauty at maturity. At UBCBG, we received this species from Hillier Nurseries, United Kingdom, under the variety *macrophyllum* (Siebold & Zuccarini) Wangerin. The plant is better placed into *Alangium platanifolium* as part of the natural variation within this species. The profile of this large shrub is vase-shaped in youth and gradually broadens with age to form a flat oval top up to 4 or 5 m high by 5 m across. It can have a short bole before dividing into several horizontal laterals that may display an undulating muscular appearance. The main branches then divide into dense clusters of twigs at the extremities. These characteristics contribute to a winter silhouette that is both distinctive and ornamental. The shrub adapts well to semishaded forest edges or as a shade-tolerant understory plant in the wild or in cultivation.

The large (16–21 cm long by 13–15 cm wide), broadly ovate yellowish green leaves resemble those of the London plane (*Platanus ×hispanica*). They are variable in shape but are generally tri-lobed, forward pointing toward the apex, with cordate bases. Often the acuminate lobes are twisted, adding to the distinctive foliar texture of this plant. In the spring the leaves unfold in the manner of hands in prayer and then turn a glorious yellow in the autumn. The flowers are white and appear in late June along the undersides of main horizontal branches often hidden by the verdant foliage. They are borne in 1- to 4-flowered cymes from the leaf axils of the previous year's growth. Each flower consists of 6 petals narrowly

93

strap-shaped and slightly twisted, forming a corolla tube at the base. Each petal reflexes to the midpoint to expose the bright yellow stamens and style. The pendulous, fleshy, egg-shaped fruit, coloured porcelain blue to dark violet, provides a stunning contrast to the golden fall leaf display.

The larger leaf variants occur widely throughout the species's native distribution, growing as an understory shrub in Japan with a host of ornamental forest dwellers such as *Cornus kousa*, *Lindera obtusiloba*, and *Helwingia japonica*. The huge natural range of this plant could provide horticulture with a broad spectrum of ornamental variation and environmental tolerance.

HARDINESS: Zones (6)7–9. *Alangium platanifolium* has not suffered from any frost damage in our garden and thrives in our West Coast environment.

CULTIVATION: It favors moist, humus-rich soil in shade or semishade, where it can provide striking contrast to the dramatic vertical forms of moisture-loving conifers native to the Pacific Northwest. The dawn redwood (*Metasequoia glyptostroboides*), vine maple (*Acer circinatum*), and skunk cabbage (*Lysichiton americanum*) make a fine trio of companions for this species. It is important to site this plant with care, so the pendulous flowers and fruit can be viewed from below. To date, no pests and diseases have afflicted this species in our garden.

PROPAGATION: The collection of fresh seed followed by a 4-month cool stratification period is the easiest method to propagate this species.

Alangium kurzii Craib
Alangiaceae (Cornaceae)   Plate 71

DISTRIBUTION: Southern and central China, India and south to the Philippines.

AN UNDERSTORY species of great beauty, *Alangium kurzii* deserves to be introduced to the West. It has astonishingly in-rolled leaf margins.

HARDINESS: Zones (6)7–9.

CULTIVATION: It favors moist, humus-rich soil in shade or semishade, where it can provide striking contrast to the dramatic vertical forms of moisture-loving conifers native to the Pacific Northwest.

PROPAGATION: The collection of fresh seed followed by a 4-month cool stratification period is the easiest method to propagate this species.

Aucuba japonica var. *borealis* Miyabe & Kudô (NA 17904)
Aucubaceae (Garryaceae)   Plates 74–75

DISTRIBUTION: Northern Japan.

THIS LITTLE-KNOWN dwarf subalpine variety of the well-known evergreen Japanese *Aucuba* hails from the "snow belt" mountains of western Honshu and Hokkaido. The shrub forms a low, undulating, ground-hugging plant that shapes itself around boulders and over fallen logs. It can grow slowly to 50 cm and thrives in moist, cool, conifer forests in the wild. The generic name *Aucuba* is a latinized Japanese word for "blue tree," *aoki*.

The diminutive, shiny, metallic, bluish green leaves are lanceolate, 5–6.5 cm long, and twisted from the midpoint to the apex. In addition, 2 pairs of teeth occur from this midpoint to the leaf tip. The plant at UBCBG is a female that produces short, cymose panicles of purple-maroon flowers from the leaf axils. In some years it produces masses of small, dark scarlet, berry-like drupes that contrast beautifully against the polished leaves. The subtle texture and form of this variety, particularly when planted in groups, increases with age.

Our plants are derived from collections made by the U.S. National Arboretum in the Gamushi National Forest, Hokkaido, from individuals growing in subalpine forests. The native environment for this plant includes a harsh winter, though heavy snowfall gives some protection to this diminutive shrub.

HARDINESS: Zones 6–7. The plant has survived temperatures to −15.6°C in the Washington, DC, area, which shows promising hardiness, particularly for moist, cool-summer climates with low nighttime temperatures. Plants have grown steadily in the David C. Lam Asian Garden since 1979, suffering no damage by temperatures as low as −13.4°C in 1990.

CULTIVATION: This attractive, compact, shade-loving shrub has thrived in the Asian Garden growing among decayed logs in the dappled shade of large conifers. It has slowly layered itself and coexists well with a fellow native groundcover, the Japanese spurge (*Pachysandra terminalis*). A number of other shade-loving shrubs also seem to associate extremely well with this plant, including our own native Cascade mahonia (*Mahonia nervosa*) and the Himalayan Christmas box (*Sarcococca hookeriana* var. *humilis*). Two other Japanese "snow belt" dwarf subalpine ecotypes that would make remarkable companions include the suckering dwarf Japanese plum yew (*Cephalotaxus harringtonii* var. *nana*), from Honshu and Hokkaido, and the chionophilous (snow-adapted) *Taxus cuspidata* var. *nana*. The latter species was collected on Mount Daisen, in southwest Honshu, by Brent Hine, in 1997. It is now well established in the E. H. Lohbrunner Alpine Garden. *Aucuba japonica* var. *borealis* has proven to be a lovely dwarf shrub free of pests and diseases and only suffering once from vole damage during one winter. The rodents ring-barked the lower stems, causing some branch dieback.

PROPAGATION: This species is difficult to grow from seed due to low viability. Cuttings taken at the greenwood stage, even in late winter, are generally successful.

Buddleja colvilei Hooker f. & Thomson
Buddlejaceae   Plate 76

DISTRIBUTION: Bhutan, Sikkim (in India).

THE GENUS *Buddleja* is named after the Reverend Adam Buddle (1660–1715), an English botanist, and the epithet *colvilei* after Sir James William Colvile (1801–1880), a Scottish lawyer and judge in Calcutta. This distinguished large shrub has often been regarded as tender in the Pacific Northwest, but has surprised me with its hardiness after a shaky start as a young plant. The species has the potential of growing with arching branches stretching to arborescent proportions. Two years ago I viewed a 15-m-high tree with an impressive trunk in San Francisco's Golden Gate Park, which revealed the shrub's robust intentions under ideal conditions and no doubt mirrored in the wild. In the David C. Lam Asian Garden, it appears to be developing into a large arborescent shrub of upright, open habit.

The striking, lance-shaped leaves (8–20 cm long by 2–6 cm wide) are tapered at both ends. As they unfurl in spring, they are covered by rusty coloured down above and felted below, which quickly falls away by midsummer to reveal finely toothed, dark green, glabrous leaves.

The flowers are perhaps the most dramatic of the genus and elicit a response of wonder by visitors to the Asian Garden. The individual bell-shaped flowers (2.5 cm across) range from rose, shades of crimson, to maroon or purple, with a white throat. Borne on long, drooping panicles (18 cm long) on last year's shoots, they appear in Vancouver during early June.

Sir Joseph Hooker had a high opinion of this species, hailing it as "the handsomest of all Himalayan shrubs." He was the first to document this plant in 1849, at 3600 m in the Sikkim Himalayas. It continues to be collected enthusiastically by today's plant explorers. Collections of a *Buddleja* with close affinities to this species in Yunnan by Keith Rushforth and Tom Hudson indicates a potential range extension of this species into western China.

HARDINESS: Zones 8–10. As a young plant in Vancouver, it was susceptible to frost damage in Zone 7b, but as it matured, it has acquired some degree of hardiness. The increasingly warm summers and gradual increase in winter temperatures here at the northern margin of its cultivation range may allow this species to be a regular subject for urban planting in the warmer areas of the Pacific Northwest.

Plate 71. The glossy leaves of *Alangium kurzii* with in-rolled margins, in the Dashahe Cathaya Reserve, Guizhou, China. Photo by Peter Wharton

Plate 72. Deep blue berries of *Alangium platanifolium* against the shrub's butter yellow fall foliage. Photo by June West

Plate 73. The flowers of *Alangium platanifolium* sporting reflexed petals, and bright yellow stamens and style. Photo by Daniel Mosquin

Plate 74. *Aucuba japonica* var. *borealis* (NA 17904), a subalpine, compact ground cover with small, terminal, purple flower clusters. Photo by June West

Above right:
Plate 75. *Aucuba japonica* var. *borealis* (NA 17904), displaying its typical glossy, black-green foliage and undulating form. Photo by Peter Wharton

Right:
Plate 76. *Buddleja colvilei*, a magnificent flowering shrub. Photo by June West

Plate 77. Semidouble, rosy pink flowers of the hardy *Camellia reticulata* (southern camellia). Photo by June West

Plate 78. The gently arching branches of *Camellia transnokoensis* spangled with fragrant, ivory flowers in the early spring. Photo by June West

Plate 79. Each flower bud scale of *Camellia transnokoensis* bears a striking red blotch. Photo by June West

Plate 80. *Cephalotaxus harringtonii* var. *drupacea* (KE 3477) displays comb-like, glaucous blue foliage and purplish red, egg-shaped fruits. Photo by June West

Plate 81. Fruits of *Cephalotaxus harringtonii* var. *drupacea* (KE 3477) in Chindo Province, South Korea. Photo by Peter Wharton

Left: Plate 82. *Clethra delavayi* (SBEC 1513) (Delavay summersweet) adorned with creamy white, lily-of-the-valley–like flowers. Photo by Peter Wharton

Above: Plate 83. The evergreen, rain-touched leaves highlight the white bracts of this woodlander, *Cornus* sp. aff. *angustata.* Photo by June West

Plate 84. *Cotoneaster glabratus* (KR 232), sporting metallic-blue evergreen leaves and dense corymbs of white flowers. Photo by June West

Plate 85. The bright red berry clusters and plum-purple stems of *Cotoneaster glabratus* (KR 232). Photo by June West

Plate 86. *Cotoneaster perpusillus*, a neat, ground-covering, pathside shrub. Photo by Peter Wharton

Plate 87. The fresh green, spring leaves and single berry of shy-flowering *Cotoneaster perpusillus*. Photo by Peter Wharton

Above right:
Plate 88. The goblet-shaped, pink flowers of *Cotoneaster splendens*. Photo by June West

Right:
Plate 89. *Cotoneaster splendens* festooned with orange-red fall berries. Photo by June West

Plate 90. The bluish green leaves of the dramatic evergreen *Daphniphyllum glaucescens*. Photo by June West

Plate 91. Pink petioles and purple stigmas of *Daphniphyllum glaucescens*, small but effective ornamental features of the shrub. Photo by June West

Above: Plate 93. The weeping branches of *Deutzia ningpoensis* 'Pink Charm'. Photo by June West

Left: Plate 92. The light pink flowers of *Deutzia ningpoensis* 'Pink Charm', borne in bold panicles. Photo by Peter Wharton

Plate 94. The charming, white to creamy yellow bells of *Enkianthus chinensis* (Guiz. 144). Photo by June West

Plate 95. The rush-like growth of drought-resistant *Ephedra gerardiana* var. *sikkimensis* thriving in sunny niches of the garden. Photo by June West

Plate 96. The striking dark reddish purple fall colour of *Euonymus carnosus* (NA 60670). Photo by June West

Plate 97. *Euonymus carnosus* (NA 60670) covered in early fall hues and developing fruits. Photo by June West

Plate 98. The pearl pink seed capsules of *Euonymus carnosus* (NA 60670) splitting to reveal black seeds and orange arils. Photo by June West

Plate 99. The late-summer, creamy white corymbs of *Euonymus carnosus* (NA 60670). Photo by June West

Plate 100. Dramatic leaf colouration of *Euonymus fortunei* 'Wolong Ghost'. Fallen seed capsules are of *Rehderodendron macrocarpum*. Photo by Peter Wharton

Plate 101. *Geum pentapetalum* (Aleutian avens), a tiny subalpine gem in the E. H. Lohbrunner Alpine Garden. Photo by June West

Plate 102. Flowers attached to the leaf midrib, a characteristic of *Helwingia japonica* (SABE 0453). Photo by June West

Above: Plate 104. Flower heads on the outer mature branchlets of *Hydrangea integrifolia.* Photo by June West

Left: Plate 103. Mature, flowering *Hydrangea integrifolia* scaling a 14-m snag. Photo by June West

Plate 105. The shade-loving *Hydrangea sikokiana* (HC 0689), with its oak-like leaves and creamy white lace caps. Photo by June West

Plate 106. The white, clustered flowers and spinose foliage of *Ilex bioritsensis.* Photo by Peter Wharton

Plate 107. *Illicium henryi* (Henry anise tree), sporting distinctive flowers and glossy green foliage. Photo courtesy of the J. C. Raulston Arboretum

Above: Plate 109. The delicate, exserted stamens and carmine-coloured corolla base of *Lonicera crassifolia* (SEH 085). Photo by June West

Above left: Plate 108. The ground-hugging gem *Lonicera crassifolia* (SEH 085), with its puckered leaves and small yellow flowers. Photo by Peter Wharton

Left: Plate 110. The stiff, upright branching and tri-lobed evergreen leaves of *Nothopanax delavayi*. Photo by June West

Plate 111. The coppery red, unfurling leaves of *Nothopanax delavayi*. Photo by June West

Plate 112. The glaucous blue leaf undersides and purplish blue drupes of *Osmanthus serrulatus*. Photo by Peter Wharton

Above: Plate 114. The floral glory of *Paeonia rockii* (Rock peony). Photo by Peter Wharton

Left: Plate 113. Clustered, creamy white flowers of *Osmanthus serrulatus*, hugging the branches. Photo by June West

Plate 115. The glossy, bluish, willowy leaves of *Persea bracteata* (YU 7835). Photo by June West

Plate 116. *Persea bracteata* (YU 7835), a handsome evergreen, offering striking contrast with the opening flowers of *Magnolia* 'Pegasus'. Photo by Peter Wharton

Plate 117. The large, whorled leaves and colourful bud bracts of *Persea ichangensis*. Photo by June West

Left: Plate 118. Distinctive, evergreen, obovate leaves with dagger-like drip-tips highlight the species *Phoebe sheareri* (NA 60723). Photo by Peter Wharton

Below left: Plate 119. The attractive leaf undersides of *Phoebe sheareri* (NA 60723). Photo by Peter Wharton

Plate 120. *Phoebe sheareri* growing among *Anthyrium filix-femina* in the garden. Photo by Peter Wharton

Plate 121. *Pileostegia viburnoides* against a Chinese character-covered rock face, in Wuyi Shan Scenic Area, Fujian Province. Photo by Peter Wharton

Plate 122. *Pittosporum truncatum* (SABE 1616), a distinctive evergreen with in-rolled leaves and long internodes. Photo by June West

Plate 123. Campanulate flowers of *Rhododendron asterochnoum* (EN 3551), with a crimson basal blotch suggesting a close kinship with *R. calophytum*. Photo by June West

Plate 124. The striking red pedicels and silvery foliage of *Rhododendron asterochnoum* (EN 3551). Photo by June West

Plate 125. *Rhododendron denudatum* (HM 1497), displaying the striped, light pink flowers and bullate leaves of this variable species. Photo by Peter Wharton

Plate 126. *Rhododendron coeloneuron* (PW 22), a close relative of *R. denudatum*, from northern Guizhou. Photo by Steve Hootman

Plate 127. The crimson-spotted, creamy yellow flowers of *Rhododendron flinckii* (KR 1442). Photo by June West

Plate 128. *Rhododendron flinckii* growing in sparse, burned *Abies* forest in northeastern Bhutan. Photo by Steve Hootman

Left: Plate 129. *Rhododendron kesangiae*, derived from wild-collected seed from Bhutan in 1988. Photo by Steve Hootman

Above: Plate 130. Floriferous *Rhododendron oligocarpum* (Guiz. 121), an annual highlight in the garden. Photo by Peter Wharton

Plate 131. Flowering *Rhododendron sinofalconeri* (SEH 229) growing at 2900 m on the Laojunshan, Yunnan Province, near the Vietnamese border. Photo by Steve Hootman

Plate 132. A wild population of *Rhododendron sinofalconeri* on the Yunnan–Vietnam border. Photo by Peter Wharton

Plate 133. *Ribes davidii* (DJHC 777), a low-growing, well-behaved ground cover with fine evergreen foliage and attractive flowers. Photo by Peter Wharton

Plate 134. The striking, vermilion red flower clusters of *Ribes davidii* (DJHC 777). Photo by June West

Plate 135. The blush pink flowers of *Rosa multibracteata* (SICH 096). Photo courtesy of Quarryhill Botanical Garden

Left: Plate 136. The clustered, delicate pink flowers and diminutive, sage green leaves of *Rosa willmottiae.* Photo by Peter Wharton

Above: Plate 137. The rounded form of *Rostrinucula dependens* (DJHC 664B) accentuated by pink-flowered, catkin-like verticillasters. Photo by June West

Left: Plate 138. The glossy in-rolled leaves and diffuse, pink flowers of *Rostrinucula dependens* (DJHC 664B). Photo by June West

Below left: Plate 139. The flowers of *Sorbaria tomentosa* var. *tomentosa*, a spectacular late summer display. Photo by June West

Plate 140. The clear yellow fall color and russet-purple fruits of the subalpine *Sorbus aronioides* (SICH 407). Photo by Peter Wharton

Plate 141. The distinctive midsummer foliage of *Sorbus aronioides* (SICH 407). Photo by June West

Plate 142. The delicate, linear, drooping, evergreen leaves and narrow, yellowish creamy racemes of *Stachyurus salicifolius*. Photo by June West

Plate 143. The racemes of *Stachyurus salicifolius* form floral curtains in the early spring. Photo by Peter Wharton

Plate 144. The rosy panicles and dew-drenched leaves of the statuesque shrub *Staphylea holocarpa* var. *rosea* (EN 3625). Photo by June West

Plate 145. The fragrant, frosty pink panicles of *Syringa sweginzowii* 'Superba'. Photo by June West

Plate 146. Graceful, cascading branches of a mature *Syringa sweginzowii* 'Superba'. Photo by June West

Plate 147. Soft pink flowers and grand foliage of a Cangshan form of *Syringa yunnanensis* (SBEC 758). Photo by June West

Left: Plate 148. Dense, colourful racemes richly adorning *Xanthoceras sorbifolium* in the early summer sun. Photo by Peter Wharton

Above: Plate 149. The cerise-coloured corolla bases, radiating stripes, and papery white sepals of *Xanthoceras sorbifolium.* Photo by Peter Wharton

CULTIVATION: When I was a young horticulture student in Britain, I often saw this half-hardy shrub trained on garden walls with southern exposures in southern and western Britain. Mature, free-standing plants make great hosts for small summer-flowering *Clematis* species within the *Connata* group, such as *Clematis rehderiana*. This species thrives in the Asian Garden in sheltered locations, in soils derived from sandy glacial till, and adapts well to tree root competition. The shrub is of the largest size for gardens with space and forest shelter in the maritime Pacific Northwest. A great companion is the seven-son flower of Zhejiang (*Heptacodium miconioides*), which grows to the same general size but its fragrant flowers appear in late summer. These shrubs can then be underplanted with ferns such as *Dryopteris wallichiana* or our own Pacific Northwest native *Polystichum munitum*, as the lower branches become shaded out and removed.

PROPAGATION: The plant is easy to raise from seed with no pretreatment. Cuttings taken between late June to late August in Vancouver root easily in a well-drained medium with mist.

Camellia reticulata Lindley
Southern camellia
Theaceae   Plate 77

DISTRIBUTION: China, in Guizhou, Sichuan, and Yunnan.

THE GENUS *Camellia* consists of over 200 species, with the main axis of diversity along the Tropic of Cancer, including the Chinese provinces of Guangdong, Guangxi, and southern Yunnan, as well as northern Vietnam. *Camellia* is named after Georg Josef Kamel (1661–1706), a Jesuit pharmacist who pioneered botany in the Philippines. In Yunnan Province, *C. reticulata* has long been revered for its longevity, evergreen habit, and spectacular yet refined flowers. Shrubs of this exquisite species have been cultivated from at least the Sui or Tang Dynasties (AD 581–906) resulting in some remarkable specimens, such as the 'Wanduocha' (10,000-flower camellia) at the Yufeng Lamasery near Lijiang. That specimen is thought to be perhaps 500 years old, flowering regularly from February to April and producing a stunning visual feast of about 4000 flowers. Peter Valder, in his superb book *The Garden Plants of China* (1999), points out that the highest concentration of "geriatric" specimens spans the ancient trade route from Myanmar through Kunming to Chengdu in Sichuan.

This shrub is truly aristocratic and given time can become a huge, domed specimen over 10.5 m tall. The leaves (2–4.5 cm long by 1–2.25 cm wide) are leathery, dark green above, generally elliptic or slightly obovate, with a distinctive acuminate tip. The name *reticulata*, meaning "net-like," describes the deeply indented, reticulate leaf veins. The leaves are glossy, with sharply serrated leaf margins—all adding to the distinctive texture of this handsome shrub. The flowers are solitary, rich rosy pink to crimson, 10 cm wide, consisting of 5 to 8 petals with yellow anthers. The flowers of *Camellia reticulata* are a truly sumptuous spectacle, often with the regal poise of tree peonies.

The species is widespread throughout western China, growing in open pine forests and scrub at 1600 to 2900 m, with an extreme high elevation of 3600 m. In Yunnan, selected forms appear to have been cultivated as far back as the eleventh century (Valder 1999), and due to the region's remote inland position, only moved slowly to the coast. From this region, Captain Richard Rawes in 1820 and Robert Fortune in 1850 introduced cultivated forms to the West. Between 1913 and 1925 George Forrest introduced wild forms of *Camellia reticulata* from the far west of Yunnan; those plants were generally noted for their small flower size or poor habit. Despite this reputation, fine semidouble or double forms are found in several areas of the natural range of *C. reticulata*, such as Tengchong, southwest Yunnan (Valder 1999).

HARDINESS: Zones 7b–8(9). The increasingly mild Vancouver winters of the last five years have certainly benefited our plants. Yet this is still a borderline hardy plant in most areas of the Pacific Northwest away from the coast. Our plants have survived −13.4°C in 1990 with minor foliage damage, giving us grounds for optimism if future climatic trends continue. Areas from the Puget Sound southward to western Oregon and into coastal California have a more favourable climate for this shrub.

CULTIVATION: We have three 4.5-m shrubs raised from seed collected at Trewithen Gardens in Cornwall, United Kingdom. The seed was collected by the owner, Michael Galsworthy, in 1984 from a group of huge shrubs grown from Forrest seed collections in northwest Yunnan. They have grown vigorously in our garden, thriving in partial shade, on glacial till that is covered in rich leaf mould. Our problem-free specimens are outstanding as they cover themselves with elegant, rosy pink, semidouble flowers every February and March. I have been impressed by the lack of pests or diseases afflicting this species.

PROPAGATION: Seeds should be collected in the fall before they harden. If they have dried out, then soaking in warm water for 24 hours can promote reasonable germination. Semimature-to-greenwood cuttings should be taken in July and planted in a peat-perlite growing medium with mist for optimal rooting success.

Camellia transnokoensis Hayata
Fragrant camellia
Theaceae   Plates 78–79

DISTRIBUTION: Taiwan, in the central mountains, at 700 to 2300 m.

THIS DELIGHTFUL *Camellia* species has a supple willow-branched form, which is initially upright then gradually assumes a more fan-shaped profile with age. In the sheltered forest glades of our garden, the plants have reached 2 m high by 1.8 m wide and are conspicuous by their open habit and delicate poise. The dainty, dark evergreen leaves (2.5 cm long by 0.5 cm wide) are oblong-lanceolate and leathery, with finely serrated leaf margins. The small subterminal flowers are apple-blossom white, pink in bud, 3 cm across, and composed of 5 petals, with white stamens and yellowish anthers. The fragrant flowers are clustered along the branches, creating an elegant fountain of graceful arching wands between late February and early March.

HARDINESS: Zones 7b–8. The original plant came to us from Nuccio Nursery, Altadena, California, in 1987, with numerous plants propagated from this mother plant. These were planted in a variety of locations throughout the David C. Lam Asian Garden, from shady, sheltered, understory sites to open, sunny spots. Since 1987 temperatures as low as −13.4°C (in December 1990) have caused considerable defoliation, bud death, and death to young plants. A sequence of mild winters beginning in the late 1990s has helped these plants reestablish themselves. Clearly a degree of caution is required when siting these plants in the landscape.

CULTIVATION: Seaside locations are ideal for this species in the Pacific Northwest, where cold-air drainage and friable, quick-draining soils with good organic content are present. Away from the coast, a semisheltered, south-facing building wall—or better still a walled garden—are good alternative sites. A cool, shaded rooting zone provided by surrounding vegetation is appreciated by this plant. This camellia is a great companion for *Arbutus unedo* 'Rubra', with both enhanced if underplanted with the glossy-leaf native Cascade mahonia (*Mahonia nervosa*).

PROPAGATION: Seeds should be collected in the fall before they harden. If they have dried out, then soaking in warm water for 24 hours can promote reasonable germination. Semimature-to-greenwood cuttings should be taken in July and planted in a peat-perlite growing medium with mist for optimal rooting success.

Cephalotaxus harringtonii var. *drupacea* (Siebold & Zuccarini) Koidzumi **Korean form** (KE 3477)
Korean plum yew
syns. *Taxus harringtonii* Knight ex J. Forbes, *Cephalotaxus drupacea* Siebold & Zuccarini
Cephalotaxaceae   Plates 80–81

DISTRIBUTION: Southern Korea, central China, and Japan.

THE SPECIFIC epithet commemorates the delightfully eccentric fourth Earl of Harrington (1780–1851) of Elvaston Castle, Derbyshire, United Kingdom. He was a great admirer of yews.

The botanical status of this fine, yew-like, shade-loving conifer is still problematic, as the Korean populations appear somewhat distinctive from their Japanese counterparts. In the wild, the Korean form of variety *drupacea* often develops into a low-domed, single-stemmed tree or large, horizontally branched shrub, 4–10 m high, finding its niche as a forest understory plant—a far cry from the small, usually under 3-m, compact shrub often associated with this variety. I have watched with deepening appreciation as these wild-derived plants have developed in the David C. Lam Asian Garden. The foliage of the Korean form is particularly ornamental, often bluish sea green, with a yellowish cast as light exposure increases.

The young stems are rugged, with distinct yellowish brown, overlapping ridges developing below each leaf petiole. Older stems often become bronze in cold winters. On older trees in the wild, the bark tends to peel and often colours to rich russet or greyish brown tones. The evergreen, linear leaves are spirally arranged and twisted at their petioles, creating 2 ranks that form a broad V shape typical of the variety *drupacea*. The leaves are 2–5 cm long by 0.4 cm wide, straight, apex acute, and often spine tipped, and are glossy, light green above, with 2 greyish white to chalk-white bands beneath.

This species is dioecious, though occasionally some individuals show a degree of confusion by changing sex after a number of years. The male flowers consist of catkins on small stalks (peduncles) that shed their pollen during March. The shorter peduncles and generally shorter leaves delineate the variety *drupacea* from the typical *Cephalotaxus harringtonii*. The prefix *cephalo* (Greek for "head") alludes to the form of the female inflorescences, which are broadly conical, consisting of a few stalked cones. Each develops into an egg-shaped, fleshy, drupe-like seed. These can grow up to 3 cm long, initially olive-green with a chalky bloom, then ripening to reddish brown to rosy red or a purplish colouration. The fruits are acrid and inedible. Interestingly, populations in southern South Korea have sweet, edible fruit; these warrant further investigation and perhaps a separate botanical status.

The UBCBG plants, under the designation KE 3477, were derived from a collection made in 1985 by the U.S. National Arboretum Expedition to South Korea, near the Ssanggye-sa Temple in Chindo Province. Seed was collected from a twin-trunked, 4-m tree growing along a tree-shaded stream. Despite the shade, this site was quite exposed, close to the sea, and often affected by salt spray during typhoons. Clearly the plant is a very tough customer, well adapted to drought, heat, and shade, and has the ability like other members of this genus to cope with intense root competition. Gardeners in deer-inhabited areas will be delighted to know that deer appear to dislike *Cephalotaxus* species.

HARDINESS: Zones 5–9. The species has a great range of temperature tolerance, from humid heat of the southern United States to the cool humidity of the Pacific Northwest.

CULTIVATION: Full exposure to the sun in the Pacific Northwest causes disfiguring chlorosis, though growth often continues to be satisfactory. In shade, vigour and colour improve. Part of the explanation for this yellowing is the summer monsoon climate of Korea. The summer cloud cover naturally depresses light levels, thus reducing the danger of chlorosis, which can afflict plants in our sunny summers. This understory species is a wonderful companion for our West Coast conifers, providing a vivid contrast to their massive, craggy stems. The Korean plum yew integrates well with shade-loving hydrangeas and snake bark maples. Since being planted in the Asian Garden in 1992, the plants have proved to be completely healthy and are now fruiting regularly.

PROPAGATION: Seeds can be collected and planted in the fall, though a 3-month cold stratification period apparently improves germination success. In general, propagation from cuttings is not an easy option. Cuttings taken in the fall can be successful after 2 to 3 months. Some evidence suggests that June cuttings develop better rooting characteristics, but a lower rooting percentage is typical (Dirr and Heuser 1987).

Clethra delavayi Franchet (SBEC 1513)
Delavay summersweet
syns. *Clethra esquirolii* H. Léveillé, *C. monostachya* Rehder & E. H. Wilson
Clethraceae   Plate 82

DISTRIBUTION: Western and southern China, at 300 to 4000 m, and northeast India, Bhutan, north Myanmar (Burma), and Vietnam.

THIS PLANT's epithet is named after the intrepid French missionary and pioneering botanist Pierre Jean-Marie Delavay (1834–1895), who documented over 1500 plants new to Western science in western China. This species is one of his most notable findings. The shrub enjoys mild maritime gardens and has a reputation for requiring exacting site requirements to succeed. These comments are still accurate even with the influx of new seed collections from China since the 1990s. Hardiness is still a major concern, so we will want to focus future collecting activities at the highest elevations in northwest Yunnan, where populations are found at up to 4000 m elevation.

This species makes a large, upright shrub to small tree 4.5–12 m high with a regular branching habit when young. Under favourable conditions it can develop into a small tree, with attractive, grey-brown, exfoliating bark. *Clethra* is derived from the Greek word *klethra*, the name for alder (*Alnus*), due to the resemblance of the leaves. The handsome, tapered, lanceolate leaves (6–15 cm long by 2.5–6 cm wide) are covered in white down that remains at maturity on the underside of the leaves. The fresh unfurling leaves in the spring can be a spectacular bronzy red. The terminal, one-sided racemes consist of clustered masses of honey-fragrant, white flowers, shaped like lily-of-the-valley, with conspicuous brown anthers and yellow stamens. The flowers jut out away from their supporting branches. Blooming takes place in late July in Vancouver.

This single, highly variable species has a widespread continuous distribution across China and adjoining countries to the south. The species is widespread in western Yunnan, from where George Forrest introduced it to Western gardens in 1913. In the wild it occurs from open fir forests to exposed limestone crags to sheltered streamside locations.

HARDINESS: Zones 7–9. In the early 1990s we lost several young plants at our garden from temperatures as low as −13.4°C, which probably sets the limit of hardiness in North America. The species appreciates cool summer nighttime temperatures, so even in the maritime Pacific Northwest and the fog belt areas of northern California, it will be restricted to cool, moist sites that are naturally buffered from the worst effects of periodic waves of arctic cold.

CULTIVATION: In cultivation our plants favour streamside locations in protected semishade. They are derived from seed collected at 3400 m by the Sino-British Expedition to the Cang Shan (Yunnan Province) in 1981 and planted out in the David C. Lam Asian Garden in 1989. Sites near to the Strait of Georgia and Puget Sound in the Pacific Northwest, particularly close to fast-moving streams near the sea, are ideal, providing optimal moisture conditions and facilitating good cold-air drainage. The shrub appears to cope well with near-neutral to slightly acidic soils but relishes soils with good organic content. In a semishaded forest environment, waterside plantings might include clusters of *Petasites japonica* var. *gigantea, Polystichum munitum, Maianthemum racemosum*, with clumps of *Pseudosasa japonica* and perhaps a dramatic background of the late-summer flowering *Eucryphia* ×*intermedia* 'Rostrevor'.

PROPAGATION: Good results can be expected from cuttings taken in June.

Cornus sp. aff. *angustata* (Chun) T. R. Dudley
Cornaceae   Plate 83

DISTRIBUTION: China, in western Hubei, eastern Sichuan, and northeast Yunnan.

THIS PLANT is an enigma, coming to us originally as *Cornus capitata* subsp. *omeiensis* 'Summer Passion' in 1995 from Piroche Plants of Pitt Meadows, British

Columbia. Unfortunately, the name is invalid—the plant is not related directly to *C. capitata* but is much closer to *C. angustata*. Despite this botanical confusion, this shrub is a graceful evergreen that has the ability, eventually, to become a large, rounded, statuesque plant of 6 to 9 m high. It appears to form a weeping tabular profile at maturity. The fine branches are pendulous at their extremities from straight horizontal branches. The glossy leaves emerge coppery red before fading to chocolate and eventually green at maturity. The leaves are leathery and narrow with broad cuneate bases (12.5–7 cm long by 6–7 cm wide). The small, globose inflorescences consist of a compact mass of tiny flowers subtended by 4 obovate bracts that are up to 3.75 cm across, white to cream, and occasionally have a faint yellow stain. Bruce Rutherford of Piroche Plants indicates that individual plants vary in flower size and colour, with even pink-stained flower forms being observed. The "flowers" of this plant resemble a typical *C. kousa*. The crimson to orange, strawberry-shaped fruits are held upright on slender peduncles, contrasting well with the evergreen leaves.

As I alluded to in my opening comments, the true identity of this plant must await molecular investigation. It is interesting to note that an unpublished monograph on Cornaceae by David Boufford, of the Cambridge Herbaria, Harvard University, Cambridge, Massachusetts, describes a broad band of variants that occur across southcentral Sichuan to western Guizhou. These seem to be intermediate between *Cornus angustata* from eastern China and the more westerly *C. capitata* from the Tibetan borderlands. Where the two populations meet, some level of hybridization, partial differentiation or speciation, is now taking place. Evidence suggests our plants are derived from seed collected in this contact zone.

HARDINESS: Zones 7b–9. We are still guarded about the hardiness of our plants, as we planted them (in 1995) in protected sites in the David C. Lam Asian Garden. They survived the winter of 2003 to 2004, when temperatures of −10°C were recorded.

CULTIVATION: This fine evergreen seems completely at home beneath the high overhead shade of large West Coast conifers, seemingly oblivious to intense root competition and drought. It also favours moderately acidic, well-drained soils with good organic content. So far it appears free of pests and diseases, including dogwood anthracnose (*Discula destructiva*), which can be a serious problem to flowering dogwoods in our cool, moist, spring climate. This fine foliage plant is a natural associate of many woodland shrubs and trees. It provides an attractive contrast in protected Pacific Northwest seaside locations with large-leaf rhododendrons or plants with dramatic branch form, such as *Aralia elata,* or other tender, stately evergreen aralids, such as *Sinopanax* or *Schefflera*.

PROPAGATION: The optimal time for terminal softwood cuttings is late June to early July in the Vancouver area. Good-sized cuttings up to 15 cm, with at least 2 leaves retained, taken from vigorous dominant terminal shoots give the best results. Wounding followed by treatment with 8000 ppm IBA in talc generally gives good results. For an excellent description of propagation techniques for *Cornus*, see Michael Dirr and Charles Heuser (1987).

Cotoneaster glabratus Rehder & E. H. Wilson (KR 232)
Rosaceae   Plates 84–85

DISTRIBUTION: China, in Guizhou, Hubei, Sichuan, and Yunnan, at 1600 to 2800 m.

THIS SHRUB is one of a host of new *Cotoneaster* introductions that have caught my attention since the reopening of China during the early 1980s. Cotoneasters, numbering over 350 species, range across Eurasia, displaying a broad spectrum of habits from ground-hugging dwarfs to huge arborescent shrubs. The obscure generic name *Cotoneaster* is derived from *Cotonem*, the quince (*Cydonia*), and *aster*, indicating an incomplete resemblance. *Cotoneaster glabratus* is a large shrub that hails from central western China, the centre of diversity for this genus. The plant in the David C. Lam Asian Garden has grown to over 6 m in 20 years, slowly developing a graceful, weeping, fountain-like profile enhanced by glossy, evergreen, fingered leaves.

The highly waxed sheen of the evergreen leaves complements the purplish maroon shoots and stems that are maintained throughout the hottest summers. The leaves are hairless with age (glabrate, as described in the specific epithet), generally oblanceolate, and tapered at both ends, ranging 5–10 cm long by 1.2–3.8 cm wide. The upper leaf surface is bright green and glabrous, with the outer margins in-rolled, contrasting dramatically with the glaucous leaf undersides and purplish stems. The flowers are white, small, and borne in dense corymbs 4 cm across on leafy shoots. The stems of the flower clusters are often suffused with purplish maroon, elegantly highlighting the white flowers. Although the reddish orange fruits are small, they are produced in substantial clusters.

HARDINESS: Zones 6–7. Our shrub has proven completely hardy since it was planted here.

CULTIVATION: Our plant is derived from seed collected by Keith Rushforth on Emei Shan, Sichuan, in 1982. Here, it grows on the south side of a forest clearing among a manmade jungle of climbers and shrubbery, perhaps similar to its native environment. It shows impressive abilities in coping with direct sun, drought, intense root competition, and resistance to fireblight (*Erwinia amylovora*)—a disease that has afflicted neighbouring rosaceous plants. The obvious ornamental qualities of this plant plus its adaptability make it a fine choice for integrating with other tall, floriferous, background shrubs or small trees (for example, *Buddleja alternifolia* or *Photinia beauverdiana* Dashahe form) or as a freestanding specimen surrounded by choice perennials, ornamental grasses, or bamboos. Of the latter, the clump-forming *Fargesia robusta* and *F. spathacea* associate particularly well. Cotoneasters generally have fallen out of favour in the Pacific Northwest, becoming victims of monocultural planting, misuse, and disease. Jeanette Fryer, who is preparing a monograph on the genus, describes them best as plants with "undervalued versatility" (1996, 146). Three other shrubby species growing in the Asian Garden derived from seed collected by Tony Schilling, Roy Lancaster, and Keith Rushforth—*Cotoneaster cavei* (Sch. 2147), *C. insolitus* (L 680), and *C. meiophyllus* (KR 4310)—are showing considerable potential for maritime areas of the Pacific Northwest.

PROPAGATION: The fruits should be collected when ripe in late October and seeds extracted. Reasonable germination can be expected if they are exposed to an acid scarification and cold stratification, then 5 to 6 months warm followed by a 3-month cold stratification. Cuttings taken in June and July root readily in sand or peat-perlite with mist.

Cotoneaster perpusillus (C. K. Schneider) Flinck & B. Hylmö
syn. *Cotoneaster horizontalis* Decaisne 'Saxatilis'
Rosaceae   Plates 86–87

DISTRIBUTION: China, in the central-western provinces.

THE GROWTH and form of this compact species, related to the well-known rock-spray cotoneaster (*Cotoneaster horizontalis*), has impressed me by its sustained vigour in the David C. Lam Asian Garden since the early 1980s. A ground-hugging species, it has maintained a dense, domed, undulating habit. A key desirable trait, compared to most ground-covering cotoneasters, is the long-term ability of this species to suppress invasive woody weeds. The typical herringbone habit of its close relative *C. horizontalis* is perceptible, though this species tends to be more compact and prostrate. Our plants appear to conform to the cultivar 'Saxatilis', which is noted for its minute leaves and shy fruiting, as described by W. J. Bean (1976).

This prostrate shrub attains a maximum height of 60 cm and often less. The oval leaves are very small, as described by the epithet *perpusillus*, 5–8 mm long, with fork-shaped stipules in their axils. Greyish brown hairs sparsely cover the leaf undersides and are more densely present on the brownish purple stems. Fall colour is often impressive, with claret-red to orange-yellow tints predominating. The flowering display is often sparse, followed by a few orange globose fruits.

Cotoneaster perpusillus was introduced into cultivation in the West from Hubei Province by E. H. Wilson in 1908. There he found it on rocky moorland, undulating around large boulders just above the tree line. The Messrs Hesse of Germany chose this shy-fruiting cultivar in 1950 for its growth habit. Our plants were obtained from the UBC Plant Operations Nursery.

HARDINESS: Zones 6–7. This plant is a very tough customer: it not only shows complete winter hardiness in our garden but also copes with heat and drought.

CULTIVATION: The normal herringbone habit of the closely related *Cotoneaster horizontalis* is less developed in this species, yet most of the ornamental features, such as small, puckered leaves and fine fall colour, are present. It has a ground-covering habit that lends itself to group planting on a grand scale, where it can form an undulating swell of dense, mounded growth. Low-growing Kurume azaleas and *Daphne tangutica*, combined with the dainty foliage of *Sorbus cashmiriana* or *S. vilmorinii*, would create a beautifully textured scene. The only drawback of this plant is susceptibility to vole damage in plague years, when these rodents gnaw the bases of stems during the winter months, leading to spotty dieback.

PROPAGATION: Seeds should have 1.5 hours of acid treatment followed by a 3- to 4-month cold stratification for reasonable germination, with some indications that *Cotoneaster* can take up to 11 months for best results (Dirr and Heuser 1987). Good rooting can be expected if reasonably firm summer cuttings are grown in a sand medium with misting.

Cotoneaster splendens Flink & B. Hylmö
Rosaceae   Plates 88–89

DISTRIBUTION: China, in Sichuan and Xizang (Tibet).

THE ORNAMENTAL hallmark of most cotoneasters is their ability to produce an abundance of colourful fruit. This lovely species from the Tibetan borderlands of China is no exception; it has been given an Award of Garden Merit by the Royal Horticultural Society. The plant is one of a group of species that are apomictic (seed is produced without normal sexual fertilization). From a practical point of view, this trait allows nurserymen to collect seed from this species and raise basically identical progeny. The species eventually makes a wide-spreading, lax-branched specimen of up to 2.7 m high and as much across. The ovate-elliptic to almost orbicular leaves (1.5–1.7 cm long by 0.8–1.2 cm wide) are covered with a whitish indumentum on their underside that overlaps at their margins, highlighting each leaf profile. The flowers, pinkish white cymes, are insignificant, but the dewdrop-shaped orange-red berries are spectacular, borne in massed clusters along the main arching lateral branches. Fall colour ranges from yellowish orange to reddish purple, providing a vivid accompaniment to the fruiting display.

This truly splendid cotoneaster was introduced to Western cultivation in 1934 by the intrepid Swedish botanist Harald (Harry) Smith (1889–1971). He documented this plant near Kangding, in central-western Sichuan on the Tibetan border, a familiar base for Western plant explorers of the previous century.

HARDINESS: Zone 6. This robust ornamental species enjoys the cool, moist climate of the Pacific Northwest. On the East Coast of the United States, the summer heat, and especially the high nighttime temperatures and humidity, make cultivation increasingly difficult, particularly in the southern states. Fireblight is a further impediment to growing *Cotoneaster* in this region.

CULTIVATION: Like so many of the highly ornamental species, such as *Cotoneaster hebephyllus* and *C. shansiensis*, this plant favours open, sunny sites with free-draining sandy soil. It is essentially a dry-lander, often pioneering disturbed habitats and enjoying the company of shrub roses, *Hypericum* species, and many sun-loving leguminous shrubs. *Cotoneaster splendens* is well named and has a well-mannered habit, even in old age, generally remaining compact and requiring a minimum of pruning.

PROPAGATION: Seeds should have 1.5 hours of acid treatment followed by a 3- to 4-month cold stratification for reasonable germination, with some indications that up to 11 months may be required in *Cotoneaster* (Dirr and Heuser 1987). Good rooting can be expected if reasonably firm cuttings are grown in a sand medium with misting.

Daphniphyllum glaucescens Blume
Daphniphyllaceae   Plates 90–91

DISTRIBUTION: Central and southern China, Taiwan, Indochina, and Indonesia.

THIS SPECIES is widespread throughout many areas of Southeast Asia. It is one of a group of evergreen shrubs that are poorly known, both horticulturally and botanically, with several species hard to separate from each other. The widespread and increasingly cultivated *Daphniphyllum himalaense* subsp. *macropodum* is well established in Western gardens, but other species of this handsome genus are now being actively sought. Under optimal wild conditions, it has the capacity to grow to an imposing 30-m tree within primary forests. It often starts life as an understory plant, then occasionally emerges as a component of the forest canopy. In the David C. Lam Asian Garden, it has developed over the past 23 years into a stately, multistemmed, restrained, arborescent shrub nearly 6 m high.

The leaves have a sea-green bloom, or glaucescence, and are just as attractive as those of the better-known *Daphniphyllum himalaense* subsp. *macropodum*, having a more oval shape with a blunt acuminate tip. In addition, the leaves are smaller (12–17 cm long by 4–6 cm wide), the leaf undersides have a light yellow bloom, and the petioles are much shorter, measuring 1–2 cm compared to 7 to 9 cm. The whorled arrangement of the terminal leaves of this species is reminiscent of rhododendrons and has from time to time fooled a number of my rhodophilic friends. The plants are dioecious, the flowers borne in sparsely flowered racemes 7–10 cm long. The individual, minute, female flowers are delightful, if you have good eyesight, consisting of a blue, glaucescent ovary and a pair of persistent, reflexed, plum-purple stigmas. The flowers are followed by blackish purple drupes, which often persist throughout the winter.

The UBCBG plant is derived from wild collections made by the Chinese Academy of Forestry in Zhejiang Province during 1980. Its vigour, fine foliage, and hardiness, particularly when young, has impressed me over the years. A Taiwanese collection made by the Royal Botanic Gardens, Kew, in 1993, identified as *Daphniphyllum glaucescens* var. *oldhamii* (ETOT 0134), has also greatly impressed me. Individuals display a spectacular range of colouration in young, unfurling leaves during the spring, ranging from crimson, purplish red, to bright orange, and fading slowly to green as the season advances.

HARDINESS: Zones 7–9. This collection has shown considerable hardiness since being planted out in the Asian Garden in 1985, surviving unscathed at –13.4°C in December 1990. The huge natural distribution of this species indicates a considerable potential for application in many moist areas of North America.

CULTIVATION: *Daphniphyllum glaucescens* copes well with full sun, but appears to relish moist shade caused by topography or forest cover. The luxuriant foliage of this species blends well with other bold, shade-loving evergreens such as large-leaf rhododendrons, *Trochodendron aralioides*, and some of the newer aralids from Asia, such as *Nothopanax delavayi* or the splendid, statuesque, though tender, *Schefflera delavayi*.

PROPAGATION: Simple stratification is all that is required to produce good germination from seed. Cuttings are best taken at the softwood to semiripe-wood stage.

Deutzia ningpoensis Rehder '**Pink Charm**'
syn. *Deutzia chunii* Hu 'Pink Charm'
Hydrangeaceae   Plates 92–93

DISTRIBUTION: China, in Anhui, Fujian, Hubei, Jiangxi, Shaanxi, and Zhejiang.

THE GENERIC name *Deutzia* acknowledges Johan van der Deutz (1743–1788), a friend and patron of the Swedish botanist Carl Thunberg, who was a botanical pioneer in East Asia. The specific epithet refers to the modern Zhejiang port of Ningbo. A large number of new introductions have been trialled in the David C. Lam Asian Garden since Western botanical expeditions recommenced visiting China in the early 1980s. Despite these exciting introductions, few have been an improvement to the extensive palette of deutzias already in cultivation in the West, with perhaps the exception of some collections from the Sino-British Expedition to the Cang Shan (SBEC) and certainly the magnificent Taiwanese *D. pulchra*. *Deutzia ningpoensis* 'Pink Charm' has impressed me greatly by its floriferous nature and vigour. The shrub grows to moderate size (2.5 m high by 3 m wide) in full sun with exquisite, light chestnut peeling bark. The younger, outer branches then assume a more graceful arching posture as they mature and flowers start to develop. This plant was raised at the famous Hillier Nurseries in the United Kingdom in 1960.

The leaves of this species are narrow, willow-like, grey on the underside, up to 4 cm long on flowering shoots and up to 14 cm long on vigorous vegetative shoots. The flowers are borne in numerous panicles up to 10 cm long along the outer pendulous branches. The bold panicles consist of cup-shaped, soft, cool pinkish mauve flowers 12 mm across in dense clusters. For sheer flower power this species is head and shoulders above other species selections. A further advantage of 'Pink Charm' is that it flowers in early July in Vancouver well after the main flowering period for deutzias. This plant also has the capacity to layer itself and form a dense self-perpetuating colony and is certainly suitable for mass planting.

HARDINESS: Zones 7–8. This species appears to be completely hardy in maritime regions of the Pacific Northwest. In the Asian Garden, it has experienced a minimum temperature of –13.4°C.

CULTIVATION: The key to success with deutzias is to maintain a yearly pruning regimen, removing a proportion of the old flowering stems immediately after flowering to encourage the growth of basal renewal shoots. The site requirements are similar to those of most deutzias: open, sunny sites with moderately fertile, quick-draining soils. In woodland or forest gardens, maintaining adequate sunlight is vital to success. This cultivar is a great choice for mass plantings in combination with shrubby hypericums and mountain hydrangeas (*Hydrangea macrophylla* subsp. *serrata*) and smaller trees that imitate the exfoliating bark of this species, such as the Korean stuartia (*Stewartia pseudocamellia*, Koreana group). Two fine wild introductions from the 1981 SBEC, *Deutzia calycosa* var. *longisepala* (SBEC 0435) and *D. compacta* (SBEC 0644) (syn. *D. hookeriana*), have also shown promise here. The former suffered some frost damage on several occasions in the 1980s (–10° to –13°C), but has since become a large, floriferous shrub.

PROPAGATION: Seed readily germinates without pretreatment. Softwood cuttings taken in June and July will give good results within 1 month.

Enkianthus chinensis Franchet (SABE 1311 and Guiz. 144)
Chinese bell flower
Ericaceae   Plate 94

DISTRIBUTION: China, in Anhui, Fujian, Guangxi, Guizhou, Hubei, Jiangxi, Sichuan, and Xinjiang, at 900 to 3600 m.

THE JAPANESE trio of *Enkianthus* species—*E. campanulatus*, *E. cernuus*, and *E. perulatus*—is well known to Western gardeners as a group of interesting year-round, deciduous companions for other ericaceous evergreens. *Enkianthus chinensis* is one of eight species in China that are, in general, less known in cultivation, with perhaps the exception of *E. deflexus* and *E. serrulatus*. *Enkianthus chinensis* is not only a denizen of the subalpine fringe, but descends into forest glades, particularly along streams. The dancing, creamy white bells of our plants must be a delight in the wild, where they often fringe boulder-edged torrents. This plant has an interesting tiered branching habit, with distinctive branch whorls plus leaves clustered at the twig terminals. The campanulate flowers ally this species botanically with the Japanese *E. campanulatus*.

In their native habitat, they develop into large shrubs or even small trees (2.5–8 m high) under favourable conditions. Stems and stalks are usually stained red. The leaves are elliptic to weakly oblong-elliptic (1.5 cm long by 1.5–2.5 cm wide), papery, cuneate at the leaf base, with an acute apex. The bluish cast to the leaf underside is also characteristic. The leaf margins are serrated, unlike those of the closely related *Enkianthus campanulatus*, which are more crenate. Our plants show considerable variation in the nature of the leaf midrib, generally impressed on the dorsal surface and raised on the leaf underside. The flowers are corymbose-racemose or corymbose, with from 8 to 12 flowers. Our plants are white-flowered forms of a species that has blooms varying from creamy white with pink veins to yellowish orange with red veins to the more typical forms with uniform dark red flowers. Flowers appear in May through early June. The slightly recurved, campanulate corollas (7–10 mm) are arranged in dense clusters. *Enkianthus* comes from the Greek *enkyos*, meaning "pregnant," and *anthos*, "a flower," together referring to the swollen appearance of each blossom. As the days shorten in October, the leaves change to a rich golden hue. Red flower forms tend to have yellow fall colour, tinted orange and red.

In April 2000, I chanced across this species in a windswept elfin forest just below the highest peak, the Huanggang Shan (2157 m), in the Wuyi Shan World Biosphere Reserve. The reserve protects some of the wildest, untouched forests of the Wuyi Shan along the Fujian–Jiangxi border. There, *Enkianthus chinensis* forms dense thickets just below montane grassland at 1800 m, with companions such as *Clethra cavalerieri* (syn. *C. delavayi*), *Eurya saxicola*, *Rhododendron fortunei*, *R. mariae* subsp. *kwangsiense*, and *Euonymus carnosus*. At that moment, though the plant itself was grey of branch and twiggy by nature, it was easy to see how the flowers and fall foliage of this species could enhance the dramatic palette of spring and fall colour that must creep across these mountainsides.

Our first collection came to us in 1992 from the 1980 Sino-American Botanical Expedition (SABE) to Shennongjia Forest District of Hubei, under SABE 1311, collected on Mount Ma at 2150 m, close to the Sichuan border. This plant has thrived along a creek in semishade, often producing rich flowering displays. In 1987, the late Jim Russell, inspired designer of the Castle Howard Arboretum in my own native Yorkshire, sent me a plant derived from seed collected by the 1985 Sino-British Expedition to the Fanjing Shan, Guizhou. That plant, under Guiz. 144, was collected at 2300 m and seems to have a more dense, twiggy habit than the SABE specimen, while both have similar flower morphology and colour. This fine, variable species clearly should join other less well-known species, such as *Enkianthus serotinus* and *E. serrulatus*, in gracing our gardens.

HARDINESS: Zones 5–7. This species seems to mimic *Enkianthus campanulatus* in general hardiness, though I expect that plants derived from seed collected from western China at highest elevations could be hardier than to Zone 5.

CULTIVATION: *Enkianthus* favors waterside locations with well-oxygenated soils to thrive. Too often the plant is sited in arid locations where it can languish. Cool, moist, root runs and sunny tops are optimal conditions for this plant. Open sunlight and space allow it to develop an attractive, tiered, branching form with age. It is a natural companion for other ericaceous plants, notably *Rhododendron*.

PROPAGATION: Likely to be similar to the requirements for allied species such as *Enkianthus campanulatus* (see Dirr and Heuser 1987). Seeds should be sown in early March on milled sphagnum and placed under mist. No pretreatment is required. Germination normally takes 2 to 3 weeks. Cuttings can be taken in late May, even without treatment, providing above-average success. Alternatively, softwood cuttings taken in July appear to root well.

Ephedra gerardiana var. *sikkimensis* Stapf
Joint fir
Ephedraceae   Plate 95

DISTRIBUTION: Nepal, Sikkim, and Bhutan, at 3700 to 5300 m.

THE GENERIC name *Ephedra* comes from the Greek name for the water plant now called *Hippuris*, which *Ephedra* very slightly resembles. *Ephedra* belongs to a botanically interesting family, consisting of approximately 40 species, which forms a superficial link between flowering plants and conifers. *Ephedra*, known as *ma huang* in the Chinese pharmacopoeia, contains ephedrine. This potent herb has been used in diverse ways ranging from weight-loss programs to the treatment of asthma.

Ephedra gerardiana has a huge distribution across central Asia, including Tibet (Xizang), Afghanistan, and India. It often occurs at extreme elevations in the wild, crouching between boulders and in crevices. The species name *gerardiana* recognizes three English brothers, Alexander, Patrick, and James Gerard, who served in the Indian Army in the late 1700s and early 1800s. *Ephedra gerardiana* var. *sikkimensis* has a local eastern Himalayan distribution and tends to grow taller than the type species. In its native haunts it grows as a prostrate ground cover no more than 5–8 cm high, while in cultivation it assumes a more shrubby habit to 60 cm high. This ground-covering shrub has a unique form and texture. Its green cylindrical stems are rush-like, hollow, and pithy when young, becoming woody at the base as the plant matures. Pairs of small, membranous, scale-like leaves (3–4 mm) clasp cylindrical stems and appear to enclose the stems and minute lateral buds at intervals, often at major nodes. These features create the loose, brush-like texture of this plant.

The stems have a waxy, soft, sage-green colour, which is in conspicuous contrast to the red, ovoid, berry-like fruits (5 mm) that, in the wild, cluster along the stems. Heavy fruiting only occurs where male and female plants are in close proximity, combined with heat and sun. The flowers are minute, light yellow, unisexual, and usually borne on separate plants. The female flowers are terminal and solitary, consisting of 3 to 4 enclosing bracts that become fleshy, red, and sweet tasting. The male flowers are generally globose in 1 to 2 groups of sessile spikes.

HARDINESS: Zones 5–8. In view of the extreme natural elevation of this plant, possibly hardier forms are in cultivation or could be introduced from the wild.

CULTIVATION: This plant is adapted to severe high-elevation alpine environments where extreme temperature fluctuations and drought are prevalent, yet it appears to adapt well to sea-level conditions if full exposure and well-drained soils are provided. In the David C. Lam Asian Garden, it has adapted well to locally dry sunny hotspots, as a fine foreground species for exposed rocky exposures, whether man-made or natural. This species is a perfect ground-covering foil for many dry-land plants. An interesting association might include sun roses (*Cistus* spp.), *Rosa* 'Max Graf', and *Rostrinucula dependens*.

PROPAGATION: The two preferred choices for growing new stock are from seed or naturally produced layers.

Euonymus carnosus W. B. Hemsley (NA 60670)
Celastraceae   Plates 96–99

DISTRIBUTION: China, in Anhui and Hunan.

ASIA IS NOTED for an extraordinary abundance of spindleberries (*Euonymus* spp.), with some of considerable horticultural merit while others have become notorious for their invasive tendencies. In fact the generic name *Euonymus*, meaning "of good name," has an ironic twist, as it refers to the poisonous qualities of the genus to animals. Despite these negatives, some species display outstanding horticultural attributes. I have been greatly impressed by the habit, foliage quality, floriferous nature, fruiting, and vigour of our plants of this species in the Vancouver area. They have grown into large, layered, arching shrubs to 4.7 m tall by 3.5 m wide.

The shimmering, glossy, leathery, elliptical leaves (6–13 cm long by 3.75–5.5 cm wide) are clustered densely along each branch. Each leaf tapers to a distinctive twisted acuminate drip-tip. The petioles, bud scales, and vertical branch lenticels are suffused with reddish purple, which further expresses itself in spectacular fall colours of shades of red, scarlet, and plum-purple. In some mild years, however, the plant remains evergreen, with leaves sometimes falling by February in typical years.

For a spindleberry, which belongs to a group not noted for showy flowers, this species can be a showstopper. Its imposing and highly ornamental blooms are borne in dense corymbs of 5 to 8 flowers atop pedicels 5–6 cm long, from mid July to early September. Individual flowers are up to 2 cm in diameter and consist of 4 separated, creamy white petals, the dorsal surfaces resembling ruckled linen. As the petals fall and the sepals wither, they reveal a pink crescent-shaped stain at their bases. The ovule is initially pyramidal, then develops into a 4-flanged, minaret-shaped seed capsule, which appears to have come straight from a cookie cutter. The fleshy (*carnosus*), 4-lobed capsule gradually turns a stunning coral pink and eventually dehisces, to expose striking orange-red seed coats or arils.

The UBCBG plants are derived from seed collected in October 1988 along a rocky boulder-strewn torrent near the famous sacred mountain of Huang Shan in Anhui Province. Peter Bristol (Holden Arboretum), Lawrence Lee (U.S. National Arboretum), and I came across a small group of plants made conspicuous by their plum-purple autumnal tints and pearl-pink seed capsules. The wild plants grew in rotten granite rubble beside a plunge pool, overhung by a tangle of ornamental companions including *Daphniphyllum glaucescens*, *Helwingia japonica*, *Ilex wilsonii*, *Magnolia cylindrica*, and *Stewartia sinensis*. All were intimately woven together with *Berchemia floribunda* and *Clematis uncinata*, with the corpulent seedpods of *Cardiocrinum cathayanum* peeking through the tangle. A second wild accession came to us from southern Hunan, collected by the Shanghai Botanical

Garden also in 1988. It has proved a fine collection as well, with perhaps a more upright habit and narrower leaves.

HARDINESS: Zones 6–9. The wide distribution of this species in China indicates a range of potential hardiness. Our plants are proving hardy. It is gratifying to see how well this species performs in the eastern United States. A recent visit to the Arnold Arboretum, Jamaica Plain, Massachusetts, confirmed the hardiness of this species to at least Zone 6.

CULTIVATION: The Huang Shan introduction (NA 60670) has constantly impressed me by its ornamentation and adaptability to tough conditions in the David C. Lam Asian Garden. It has responded to 2 consecutive years of drought and heat with vigour and wonderful flower displays. If provided a sunny well-drained site, it should perform well throughout many parts of western and eastern North America. No sign of seedling volunteers or disease has been recorded in Vancouver; however, vigilance must be maintained for volunteers in the warmer regions of the United States, in the southeast and the moister areas of California and Oregon. Great caution should be observed when introducing any spindleberry to North America. Immediate removal of the plant is imperative if aggressive seeding is observed, especially into natural areas.

PROPAGATION: Cuttings taken in June to August readily root in a peat-perlite medium with mist.

Euonymus fortunei (Turczaninow) Handel-Mazzetti '**Wolong Ghost**'
   (DJHC 691)
Wintercreeper euonymus
Celastraceae   Plate 100

DISTRIBUTION: Central and western China, Korea, and Japan including the Ryukyu Islands.

THIS SPINDLEBERRY is named after Robert Fortune (1812–1888), who was in the vanguard of British plant collectors in the Far East. He did much to popularize and introduce Asian plants to Western horticulture, including the winter favorite, *Jasminum nudiflorum*, and his "queen of primroses," *Primula japonica*.

Many variegated cultivars of wintercreeper euonymus are well known across North America as landscapers' staple ground covers. Unfortunately too many plantings are paying the price of monocultural overuse, pests, and disease. Yet there are interesting and attractive new introductions that have great appeal if used judiciously. Dan Hinkley has given us a delightful delicate, narrow-leaf, white-variegated form of this species from Sichuan. This quality plant is already illuminating the deep, leafy shade beneath a specimen of *Rehderodendron macrocarpum* in the David C. Lam Asian Garden. *Euonymus fortunei* has an enormous range throughout China, Korea, and extending to Japan, and the Ryukyu Islands, often displaying a bewildering array of morphological forms—ground cover, shrub, or climber—depending on the character and opportunities of the natural environment.

Euonymus fortunei is generally evergreen in cultivation, though often losing a considerable proportion of its leaves in late fall or after a severe winter. Leaves are notoriously variable, generally elliptic up to 4 to 6 cm long by 2.5–3 cm wide, opposite, with crenate-serrate margins. Leaves often have a dark bluish green colour that is striking with the silvery leaf veins. The stems are trailing, often developing aerial rootlets at intervals, which then enables them to climb into trees or ascend cliffs. The leaf morphology changes from the juvenile to the adult flowering state, in a manner similar to ivy (*Hedera helix*). The flowers are underwhelming, produced in axillary cymes of 4 flowers (6.5 mm) and of an unremarkable greenish white. Attractive clusters of pinkish to reddish capsules appear in the early fall; these split to expose bright orange, fleshy arils that persist into late November.

I have personally seen this species on a number of occasions in China, but the first prize goes to a huge, cliff-hugging specimen on Yu Ping Lo (Jade Screen Peak) in the sacred Huang Shan of southern Anhui Province. This specimen scaled a splintering granite cliff up to 12 m, with the stiff outer branchlets adorned with heavy crops of rosy red fruits. The plant agrees with *Euonymus fortunei* var. *radicans* (Siebold ex Miquel) Rehder, which in general has a more diminutive leaf morphology but has the same vertical ability of the type.

Euonymus fortunei 'Kewensis' is another elfin favourite of mine, with dimple-sized leaves and a ground-hugging, prostrate form that delicately clambers over shady rock exposures.

The ability of this species to assume different forms, thrive in full sun or deep shade, and be pH-adaptable ensures success in the wild. An example of this versatility is the handsome *Euonymus fortunei* 'Wolong Ghost' under Dan Hinkley's collection (DJHC 691) from near the Wolong Nature Reserve in Sichuan, famous for its giant pandas. This cultivar is a delight, with dark green, 7- to 8-cm long leaves that are etched startlingly white along the central and secondary veins.

HARDINESS: Zones (4)5–9. The huge range of this plant with diverse provenances gives horticulture a breadth of potential cold-hardy genotypes. In general, this species is very adaptable for both eastern and western North America.

CULTIVATION: This species and its forms are first-class plants for adorning hard structures or arbors, with the capacity to cope with concrete-derived leachate and dry, exposed environments. Admittedly, *Euonymus fortunei* is regarded as a serious woody invasive in eastern and central western regions of the United States. Caution and vigilance should be applied when utilizing this species anywhere in North America. In southwest British Columbia, however, no serious signs of seedling invasion have yet been observed.

PROPAGATION: Cuttings taken in June to August root in a peat-perlite medium with mist.

Geum pentapetalum Makino
Aleutian avens, Chinguruma
syn. *Sieversia pentapetala* Greene
Rosaceae   Plate 101

DISTRIBUTION: Eastern Asia (Japan, Kamchatka, and the Kurile Islands), and the Aleutian Islands (United States).

THIS FASCINATING little subshrub of the eastern Pacific maritime mountains has a curious growth form. The base is woody with herbaceous outer branchlets creating a brown pile of twining stems to 10 cm during the winter months. The species is subalpine and often buried in heavy snows for most of the winter. The Japanese common name, *chinguruma*, is a corruption of the word *chigoguruma*, or "cart for a little child," which alludes to the seed heads having the shape of a pinwheel.

This plant has small, dark, glossy green leaves. Each delicate compound leaf consists of 3 forward-pointing, deeply cut, obovate leaflets. The flowers consist of 5 white petals—therefore the epithet *pentapetalum*—combined with golden stamens that are held poised attractively above the foliage. Flowering occurs in early summer, followed by plumose seed heads, similar to those of *Anemone* or *Pulsatilla*. *Geum* is the classical name for avens, or herb bennet.

This species thrives on rocky subalpine grasslands, often favouring volcanic rock crevices, where seepage from melting snow patches keeps the soil moist and cool throughout the growing season. This plant has an interesting northern circum-Pacific distribution from Japan to the Aleutian Islands.

HARDINESS: Zones 6–7b. We have had no problems with hardiness at close to sea level at UBCBG, yet this is a plant that is often snow covered for perhaps 6 to 7 months of the year in the wild.

CULTIVATION: At lower elevations, the plant thrives in a rock-garden setting that faces northeast. Recommended soils are cool and well drained and allow seepage of water through the root zone. Soils are generally rich in grit and organic matter in the wild and should be replicated under cultivation. Raised beds or other locations that allow quick drainage are also ideal. Attractive associations utilizing this plant could include *Dryas octopetala*, or mountain avens, and some of the dwarf, moisture-loving, alpine willows such as *Salix nakamurana* var. *yezoalpina* or our native, subalpine, netted willow *S. reticulata*. The greatest challenge for the grower at sea level is to mimic the plant's natural growing conditions, particularly reducing sudden temperature extremes during the winter when the plant normally would be protected under snow.

PROPAGATION: Fresh seed germinates readily with normal stratification.

Helwingia japonica F. Dietrich (SABE 0453)
Helwingiaceae   Plate 102

DISTRIBUTION: Central and western China, Bhutan, Myanmar (Burma), Taiwan, and Japan.

THE GENUS *Helwingia* commemorates the German pastor, G. A. Helwing (1666–1748), who worked on the flora of Prussia and is perhaps best known for his work on the genus *Pulsatilla*. Sometimes regarded as no more than a botanical curiosity, *Helwingia* has charm and a quiet ornamental character. Our plant, with its leaves of appealingly delicate form and texture, complements more demonstrative ornamentals.

We have been growing *Helwingia japonica* for many years derived from seed collected by the 1980 Sino-American Botanical Expedition to Hubei (SABE). The species has a huge distribution in Asia, with many regional variants. At a cursory glance, the plant resembles a deciduous *Aucuba* with foliage and stem morphology closely approximating this genus. Our plant has grown into a handsome, arching shrub (4.5 m high by 5 m wide) with a distinctively open centre revealing a cluster of yellowish green stems, reminiscent of *Leycesteria formosa*. The bright green, tapered leaves are ovate to slightly lanceolate on mature foliage (3.5–9 cm long by 2–6 cm wide), while juvenile leaves tend to be broadly elliptic. The petioles are up to 5 cm long and along with the midvein are often stained pale pink. The bristle-like teeth at the margins and hair-like stipules are distinctive, but the most curious feature is the epiphyllous flowers (borne upon pedicels attached to the leaf midrib). The fall colour is a strange purplish grey. The flowers are monoecious, an insignificant greenish white, with male and female flowers arranged in small umbels in the centre of leaves. The fruits are ovoid, black, and rarely produced in profusion.

Our plants were originally labeled as *Helwingia chinensis* under SABE 0453, collected near Mucheng, in the Shennongjia Forest District of Hubei Province. Perhaps the most remarkable concentration of *Helwingia* species is in southcentral Sichuan, particularly in the vicinity of Emei Shan, where four species and many local variants can be found, including *H. chinensis*, *H. himalaica*, *H. japonica*, and *H. omeiensis*.

HARDINESS: Zones 6–8. This species has an enormous natural range, so seed collections from the lower mountains of southern China are perhaps ideal for the warmer areas of the United States on both coasts. Our collections from Hubei appear ideal for the Pacific Northwest, and show no frost damage in 20 years.

CULTIVATION: This ornamental is understated and ideal for shady, moist, forest gardens as part of a shrubby understory. The delicate foliage and open habit provide a delightful counterpoint to other shade-loving evergreens so often used with Pacific Northwest conifers. The combination of this refined shrub with other shade lovers, including *Aucuba japonica* 'Salicifolia', *Cephalotaxus fortunei*, *Lindera obtusiloba*, bamboo, and a foreground of *Sarcococca hookeriana* var. *humilis*, could work well. As the shrub ages, some of the older stems die and require removal, which encourages the growth of renewal shoots from the base by providing space and additional light.

PROPAGATION: This species spreads slowly by lateral root suckers (ramets). These can be used for propagation if a sharp spade or knife is used to sever the ramets for replanting. Seed from ripe black fruits may be sown. A normal 4- to 5-month cool stratification period should result in good germination.

Hydrangea integrifolia Hayata
Hydrangeaceae   Plates 103–104

DISTRIBUTION: Taiwan, the Philippines.

THE NAME *Hydrangea* comes from the Greek *hydro*, meaning "water," and *angeion*, meaning "a vessel," in reference to the dehiscent seed capsules, which are shaped like a Grecian water jar. This remarkable, evergreen, climbing hydrangea belongs to the distinctive section *Cornidia* of the genus. Curiously, members of this section have a disjunct distribution, with *H. integrifolia* being the only Asiatic representative of about a dozen species from Central and South America. In the wilds of Taiwan's central mountains, the Chungyang Shan, this climber attains a massive size, ascending large forest trees or cliff faces. As a juvenile it often scrambles in the dark as a ground cover searching for a ladder to the light.

The dark green, evergreen leaves of mature, flowering specimens are generally oblong-elliptic (7–22 cm long by 3.5–8 cm wide), leathery, and glabrous, with pink petioles. The specific epithet *integrifolia* refers to the entire, or uncut, leaves. The juvenile leaves on young plants and those on actively ascending branches are smaller and more lanceolate. The young stems are covered in dense, coarse, bristly, reddish purple hairs, each crowned with a white, stellate cluster of filaments. Actively growing spring shoots and leaves are often coloured spectacularly, ranging from plum-purple to bright crimson red, depending on individual plant variation. Aerial roots allow this plant to clasp any vertical surface, and they become dense and prominent as they mature in the manner of *Hedera*. Individual plants can ascend to at least 30 m if given a forest tree or cliff to scale.

This spectacular vine produces greenish white flower buds that gradually expand to the size of golf balls. The flower buds against the dark green foliage gives a polka-dot appearance to mature flowering specimens. Inflorescences are terminal, with prominent outer, rounded, floral bracts typical of the section *Cornidia*. Individual flower buds only partially open by late June and reveal a mass of creamy white stamens that protrude just beyond the cup-shaped blooms. In flower this species elicits comments of wonder from the general public in our garden.

HARDINESS: Zones 7b–9. We have grown plants from two separate introductions, one by Pierre Piroche, from Alishan at 2400 m in 1986, and the other in 1993 by the Taiwanese Forestry Research Institute, from the Nantou–Hualien County boundary at 2720 m. We lost a few plants of the former collection in December 1990 (at −13.4°C), but since then our plants have suffered no damage.

CULTIVATION: This plant relishes montane cloud-forest environments, with abundant moisture and shelter. Like *Hedera*, it is adapted to forest floor environments, where shade and intense root competition are the norm. It is relatively easy to establish this plant in our West Coast rain forests, close to potential large tree hosts, snags, or natural cliffs. Careful watering in the first year is important when young plants are sited near trunks of large trees, such as western red cedar (*Thuja plicata*), which divert natural rainfall to the dripline zone. A forested ravine with rock faces, plunge-pools, combined with dramatic evergreens and snake bark maples would meet the needs of this exceptional climber. Minor problems can occur in youth, including slugs browsing on soft, expanding shoots and, occasionally, weevil damage to mature foliage below about 3 m.

PROPAGATION: The same prescriptions described for *Pileostegia viburnoides* should apply to this species.

Hydrangea sikokiana Maximowicz (HC 0689)

Hydrangeaceae   Plate 105

DISTRIBUTION: Japan, in western Honshu, Shikoku, and Kyushu.

JAPAN IS regarded by many as being one of the few Asian countries that has been thoroughly scoured by plant hunters, from Richard Oldham (1837–1864) and Charles Maries (1851–1902) to contemporary collectors such as John Creech,

Barry Yinger, and Mikinori Ogisu. This is far from the truth, which brings us to this remarkable shrub, which has a personality of delicate charm although closely related to the impressive, husky North American *Hydrangea quercifolia*. It took the persistence of a quartet of modern-day plant explorers, Dan Hinkley, Bleddyn and Sue Wynn-Jones, and Daryl Probst, to introduce this handsome woodlander to the West in 1997. Hydrangeas are extremely well represented in the Japanese flora, with up to 10 species recorded and numerous subspecies and varieties, not to mention the many cultivars, some of which still need to be introduced.

In the wild this shrub is a small, deciduous, wide-reaching, open plant 1–1.6 m high. The branches are stout and rounded in youth, before starting to exfoliate as they mature. The imposing, deeply scalloped leaves can be massive, generally broadly ovate, 10–20 cm long by 7–15 cm wide, sharply serrate, but weakly incised toward the apex. The leaves are coarsely felted on both sides, especially on the lateral veins. The light green, upper leaf surface is often mottled yellow, while the underside is uniformly paler.

Under optimal conditions, leaves can attain gigantic sizes up to 38 cm long, reminiscent of canoe paddles. The petioles, or "paddle handles," can also be substantial, 5–15 cm long. The flowers follow suit in concert with the scale of their leaves. The inflorescence consists of a large cyme that is up to 30 cm across, hirsute, with large marginal sterile bracts. The fertile and sterile florets are white and appear in midsummer as ghostly orbs in the shady gloom of a wild or cultivated forest setting. In winter, old flower heads are often embellished by snow lodging on their flat surfaces. As the shrubs mature, exfoliating bark gradually becomes an increasingly noticeable ornamentation.

HARDINESS: Zones 7–8. A Christmas 2003 cold spell (to −10°C) has had no effect on our plants and gives us hope that this species will become a fine new introduction for West Coast gardens.

CULTIVATION: The UBCBG plants are derived from seed collected from the cool-temperate highlands of Honshu, on the Kii Peninsula, growing with a wealth of hydrangeas along small, shaded streams. We should try to mimic those conditions in our own gardens if we are to grow this species successfully. This plant is a good choice to use with other large-leaf architectural hydrangeas, such as *Hydrangea aspera*, with velvety, villose leaves, and the stunning *H. sargentiana*, with thick, felted stems. Other companions for *H. sikokiana* for streamside situations could include the glorious skunk cabbage (*Lysichiton americanum*) and devil's club (*Oplopanax horridus*). These two natives are wonderfully sinister denizens of the gloom—the former is a native starvation food, and the foliage of the latter should be kept away from skin contact. The greatest drawback of this species is that it is prone to being victimized in youth by slugs, and so some kind of control measure is required.

PROPAGATION: The following recommendations are applicable to *Hydrangea quercifolia*, the North American counterpart of this species. Cuttings should be taken in mid to late July and planted in a well-drained medium with mist.

Ilex bioritsensis Hayata
syns. *Ilex diplosperma* S. Y. Hu, *I. pernyi* var. *veitchii* Bean
Aquifoliaceae   Plate 106

DISTRIBUTION: China, in Guizhou, Hubei, Sichuan, and Yunnan, and in northeast India, in Manipur, as well as in Taiwan.

THIS NOBLE holly has been a nomenclature nightmare for those trying to pin down its true identity. We have grown a plant from the Hillier Nurseries for many years under *Ilex pernyi* var. *veitchii*; it is *I. bioritsensis*. The species is noted for its immense vigour and high-gloss, spinose foliage reminiscent of the more diminutive *I. pernyi*. In 1908, T. Kawakami and U. Mori originally collected *I. bioritsensis* at Bioritsu, Taiwan.

Horticulturally, this plant is a giant version of *Ilex pernyi* and can attain single- or multiple-stemmed tree status under favourable conditions or given time. As a young plant, it can form a tight, triangular spire and maintain this form well into middle age before the top rounds over and the general profile becomes irregular. The rigid branches are clothed in dense, glossy foliage of exquisite texture. The leaves (3.8–5.7 cm long by 2.5–3.8 cm wide) are leathery and weakly truncate to cordate at the base, with a broad, triangular, spinose apex. The leaves have undulate, reflexed margins with between 3 to 6 spines on each margin.

The flower buds are brownish purple, grouped in clusters in the leaf axils. The flowers are pale yellow and 4-merous, often clustered densely along the branches of mature specimens growing in open situations. Fruits are often bright red and bunched tightly to the stems. The fruits are highly ornamental, especially as the plant ages and the leaves decrease in size.

This shrub is one of the largest size and not for the faint of heart. Its armature is substantial in youth, like many evergreen *Ilex* species. The spines become less prominent on mature flowering plants, but often remain on lower branches.

Ilex georgeii is a desirable relative of *I. bioritsensis* from Yunnan, northern Myanmar (Burma), and Manipur in India. This species could be just a western expression of *I. bioritsensis*, but the extremely dense, narrow, spinose foliage and profuse berry production witnessed in cultivated plants in Kunming, Yunnan, warrants further investigation.

HARDINESS: Zones 6–9. *Ilex bioritsensis* appears hardy in many areas of the maritime Pacific Northwest and the eastern and southeastern United States. In the David C. Lam Asian Garden, it has been reliably hardy and extremely heat tolerant.

CULTIVATION: This tough, resilient evergreen copes with a wide range of challenging sites from full sun to deep shade and a variety of soil types except those with poor drainage. This species is versatile and can be used for windbreaks, hedges, and as a fine understory specimen. The only pest that has occasionally caused minor damage is the holly budmoth (*Rhopobota naevana ilicifoliana*), a coastal Pacific Northwest pest. The adult female lays eggs on the underside of leaves in late summer, and the eggs hatch in the following late April to early May, with the larvae migrating to the terminal shoots. Here the young leaves are webbed together with silk before feeding commences. A program of monitoring in mid April is required, to synchronize treatment with this insect's window of vulnerability when it is on the move. We have found natural predation particularly by birds generally keeps this problem under control. If the problem persists over a period of years, then spraying with Bt (*Bacillus thuringiensis*) at the motile larval stage works well.

PROPAGATION: Propagation of *Ilex* is often difficult from seed, sometimes requiring up to 3 years before germination occurs. In our experience, terminal cuttings with 2 retained leaves should be taken during September through early March and placed in a sand-peat mix with bottom heat and mist.

Illicium henryi Diels
Henry anise tree
Illiciaceae   Plate 107

DISTRIBUTION: China, throughout eastern and western central regions.

THE ANISE FAMILY consists of a single genus, *Illicium*, and approximately 40 species, 28 of which are native to China. Several species, notably the Chinese star anise (*I. verum*), are grown commercially for their essential oils often used for flavoring. The generic name *Illicium* refers to the alluring, aromatic oils that are found in all parts of these plants. Some species are grown just for their ornamental attributes, such as *I. henryi*.

The specific epithet recognizes Augustine Henry (1857–1930), who was a man of many talents, including physician, plant collector, botanist, and author, who served with the Imperial Chinese Maritime Customs Service at Yichang (Ichang) on the Yangtze River from 1889 to 1892. He is best known for his valuable assistance to E. H. Wilson in his early explorations. This species in the wild is often a low-growing, spreading shrub beneath large forest trees, in thickets or swamps.

If grown in the open or allowed to emerge into the light, it gradually becomes a dense conical to pyramidal shrub or even a small tree (3–8 m, and possibly to 12 m).

The alternate leaves are dark green, leathery, glossy, oblanceolate to long lanceolate (6–18 cm long by 1.2–5 cm wide), displaying cuneate bases and long acuminate apices. The faintly pink petioles and buds of the current season's foliage hint at the colour of the flowers, which are variable in form, either axillary, subterminal, solitary, or in fascicles of 2 or 3. The flowers appear in early April and are usually up to 2.5 cm across, consisting of 10 to 25 pink to dark red tepals. The fruit is the characteristic star-shaped aggregate of follicles (2.5 cm) similar to the frequently grown Japanese anise tree (*Illicium anisatum*).

Two species that are worth trying in the Pacific Northwest include *Illicium lanceolatum*, from east central China, and *I. simonsii*, which is widespread throughout western China. The latter has impressed me greatly on a number of visits to Quarryhill Botanical Garden, in California, famous for its remarkable wild-derived collections of Asian plants. A small hillside grove of this species, collected under HM 1400 from Sichuan, displays to perfection several outstanding attributes of this species, including a compact habit and glorious sage blue foliage.

HARDINESS: Zones 7–9. Our plant came to us in 1988 from Camellia Forest Nursery under their collection number, 85C–24. It survived −13.4°C in December 1990 with no damage recorded, and is capable of thriving in many areas west of the Cascade Mountains in the Pacific Northwest. The greater summer warmth and rainfall of the southeastern United States certainly enhances the vigour of this plant.

CULTIVATION: Our plant was initially sited in deep shade in a protected part of the David C. Lam Asian Garden 20 years ago; we had no idea of its hardiness at that time. It has thrived despite difficult conditions, but awaits release into full sun and more abundant moisture. The broad spectrum of tolerance of this plant is impressive, ranging from deep forest shade, to swampy sites, to full sun, making it a useful shrub for diverse landscapes and functions. A lovely woodland plant in its own right, it can easily be associated with familiar camellias, magnolias, and rhododendrons. An intriguing combination for shady forest conditions would be with genera such as *Daphniphyllum*, *Mahonia*, *Pittosporum*, and with many new *Epimedium* species that have been introduced from China since the 1980s.

PROPAGATION: Seed germinates readily with no pretreatment. Firm wood cuttings can be taken between July and September. Rooting may take up to 2 months and success is often high.

Lonicera crassifolia Batalin (SEH 085)
Caprifoliaceae   Plates 108–109

DISTRIBUTION: China, in western and northern Guizhou, southwest Hubei, northwest Hunan, southern Sichuan, and Yunnan, at 900 to 2300 m.

STEVE HOOTMAN of the Rhododendron Species Botanical Garden in Washington State collected this diminutive gem on one of many exploration trips to China scouring the terrain for new rhododendron species. This shrubby ground cover is one of the most beautiful creeping shrublets to be introduced into cultivation in the last few years. Augustine Henry first recorded this undersized plant in 1889 on the Wa Shan in southern Sichuan, close to the sacred mountain of Emei Shan (Mount Omei). The shrubby honeysuckles are in general poorly known, particularly in the maritime areas of the Pacific Northwest, with perhaps the exception of *Lonicera pileata* and some of the winter-flowering species like *L. fragrantissima*. This distinctive species fills a useful, midsummer-flowering, ground-covering niche for our gardens.

In cultivation it forms a low, dense, prostrate mound up to 16 cm high, and it is capable of covering several square metres over time. The tiny, fleshy, puckered, ovate leaves are arranged in a pectinate manner. The specific name *crassifolia* refers to the thick, fleshy leaves. The flowers are white, turning pale yellow as they age, with the base of the corolla tube remaining a delicate carmine colour. These attractive bicoloured flowers appear in early May, followed by equally lovely blue berries in the fall.

The UBCBG plants are derived from a wild collection made by Steve Hootman during September 1995 from a deep, shaded ravine in the Jinping Shan of southern Sichuan at 2895 m. As Steve described the native environment of this plant:

> This is a karst range with steep slopes and lots of waterfalls and a rich acid-loving flora. . . . This is an incredibly rich area with masses of different rhodies and other goodies like *Berneuxia thibetica* and *Ribes davidii*. The plant was growing as a creeping prostrate 'vinous' subshrub on the forest floor in deep shade, and at first I thought it was a *Mitchella*.

It also appears to love growing over moist, moss-covered rocks and fallen logs in the company of a palette of epiphytic plants including *Agapetes*, *Gaultheria*, *Pleione*, *Rhododendron*, and *Vaccinium*. This certainly suggests that these genera are potentially excellent associates for this plant in cultivation. *Lonicera crassifolia* also forms a beautiful blend of ornamental characteristics when planted with *Ribes davidii*.

HARDINESS: Zones (6)7–8. The jury is still out on this species, but from my observations of plants growing at Heronswood Nursery and the Rhododendron Species Botanical Garden, both in Washington State, a rating of Zones 7 to 8 is not unreasonable. Extreme high-elevation forms may push the rating close to Zone 6, particularly where regular winter snow cover occurs.

CULTIVATION: *Lonicera crassifolia* is already showing great potential as a flowering ground cover for open to densely shaded sites. It appreciates moisture yet seems to have a degree of drought tolerance. Soils should be free draining, with plants showing increasing vigour with higher organic content in the substrate. Already the creeping shrublet appears to appeal to the general public as a container plant. Better flowering should be expected on more open sites.

PROPAGATION: Nodal tip cuttings or double leaf bud cuttings should be taken between June and August.

Nothopanax delavayi (Franchet) Harms (EDHCH 97321)
Delavay false ginseng
syn. *Metapanax delavayi* Franchet
Araliaceae   Plates 110–111

DISTRIBUTION: China, in Guizhou and Yunnan, at 1400 to 2500 m.

ARALIACEAE is a huge, diverse family that has given horticulture many fine ornamental plants with attractive foliage. Recent introductions have sparked much interest—many are grandly tropical in appearance, with some showing a surprising degree of cold hardiness. *Nothopanax* is derived from the Greek *nothos*, meaning "false," and *panax*, the generic name for ginseng. The specific epithet refers to the French missionary and botanist Jean Marie Delavay (1834–1895).

   This large, upright shrub has a short bole, which divides into several strongly ascending stems. The glossy evergreen leaves are polymorphic, from simple to bilobed, with the mature upper and outer leaves tending to be trilobed, with each lobe lanceolate. The leaves are up to 15 cm long by 12 cm wide, with pendulous, stretching petioles up to 14 cm long, giving a refined appearance. The young unfurling leaves glow coppery red before assuming their mature green foliage in early summer. Our plant was collected by Eric Hammond, of Heronswood Nursery in Washington State, near Boaxing, Sichuan, at an elevation of 1676 m, indicating a degree of hardiness. The flowers have large, umbellate panicles consisting of greyish white flowers, before developing into clusters of attractive black fruits.

HARDINESS: Zones 7–9. It is difficult to estimate hardiness of this plant at UBC, as we have had a sequence of mild winters since it was planted out in the David C. Lam Asian Garden in 2000. Climatological reports from the J. C. Raulston

Arboretum, Raleigh, North Carolina, and the Sarah P. Duke Gardens, Durham, North Carolina, indicate it could be reliable in Zone 7b.

CULTIVATION: This species appears to thrive well west of the Cascade Mountains in the Pacific Northwest. If grown in full sun, it readily forms a compact, upright specimen up to 5 m, while in deep shade with root competition a height of 2 to 3 m is more likely. Deep moist soils rich in surface organic matter are preferable, though it appears to tolerate more stressful sites. This plant is a fine subject for shady forest gardens requiring shrubs of architectural merit. It is a good companion for its aralid relative *Aralia elata*, the Japanese angelica tree. It also looks at home with massed waterside plants, such as evergreen ferns, bold herbaceous plants such as *Lysichiton* species, or combined with woody shade lovers like *Acer circinatum*, *A. japonicum*, *Lindera obtusiloba*, and bamboos. Other relatively hardy, large, evergreen shrubs within Araliaceae that deserve to be better known include *Schefflera taiwaniana* and the glorious *Sinopanax formosana* that flaunts sage green, rusty chocolate indumented leaves.

PROPAGATION: Typical of many aralids, seed from only black, ripened fruits should be sown. Inhibitors within unripened fruits can reduce germination levels significantly. No pretreatment is necessary but bottom heat at 26°C enhances germination.

Osmanthus serrulatus Rehder
Oleaceae   Plates 112–113

DISTRIBUTION: China, in Fujian, Guangxi, and Sichuan, at 700 to 2000 m.

A GROUP of evergreen shrubs to small trees closely related to *Olea*, the olives, and *Syringa*, the lilacs, *Osmanthus* is often given the vernacular name "sweet olive." In China there are 23 species, among which the best known is *Osmanthus fragrans*, a shrub revered for its association with noble qualities (Valder 1999) and practical utility, from flavoring wine to making perfume or medicines. *Osmanthus serrulatus* may not have the charisma of *O. fragrans*, but for 22 years it has grown with great vigour and excellent health in the David C. Lam Asian Garden. Our specimen has been slowly drawn upward over the years by the increasing shade of several surrounding 30-m-tall grand firs (*Abies grandis*). Despite this context, it has remained a most ornamental species, coping with drought and shade and flowering well.

In the wild and in long-term cultivation, *Osmanthus serrulatus* is capable of reaching 8 m. Cultivated plants in full sun are likely to attain 4 m, gradually forming a spreading dome. The lustrous, dark green leaves with a paler underside range from narrowly obovate to elliptic (6–14 cm long by 2–4.4 cm wide). The leaf margins, midrib, and subsidiary reticulate venation are stained chartreuse. The serrate leaf margins are conspicuous, with up to 25 pairs of tiny piranha-like teeth. These spinose leaves gradually become unarmed as the shrub matures, in a manner similar to *O. armatus* and some hollies (*Ilex*).

The word *Osmanthus* is derived from Greek *osma*, meaning "fragrant," and *anthos*, meaning "a flower," acknowledging these generic features. The flowers are borne at the leaf axils, in cymes consisting of 4 to 6 fragrant, pure white flowers. Clustered along the main stems in May, they contrast beautifully with the plant's black-green leathery leaves. In sunny, hot years, fine crops of peanut-size, purplish blue drupes are borne on 3-cm-long pedicels.

HARDINESS: Zones 7–9. This shrub has remained free of frost injury for 22 years and may be hardier than indicated.

CULTIVATION: This shrub is a tough evergreen for background screening and is also able to fend for itself in challenging landscapes. It associates well in drought conditions with subjects like *Arbutus unedo* and *Escallonia* species. The species has a strong constitution and remains free of pests and diseases. *Osmanthus serrulatus* joins a number of desirable species from China, including *O. armatus*, *O. delavayi*, *O. henryi*, and the incomparable but tender large-leaf species, *O. yunnanensis*. The UBCBG plant originally came to us in 1982 from Hillier Nurseries, Winchester, England.

PROPAGATION: Firm cuttings taken in late summer seem to give the best results.

Paeonia rockii (S. G. Haw & L. A. Lauener) T. Hong & J. J. Li
Rock paeony
syn. *Paeonia suffruticosa* subsp. *rockii* S. G. Haw & L. A. Lauener
Paeoniaceae   Plate 114

DISTRIBUTION: China, in southern Gansu, western Henan, western Hubei, and southcentral Shaanxi, at 1100 to 2800 m.

THE GENUS *Paeonia* comes from the Greek name *Paeon*, the physician of the gods and the discoverer of medicinal properties. This shrub has great popular distinction for its bold, distinctive foliage, vigorous form, and exquisite flowers of grand proportions. The tree peonies in general and this species in particular have been especially esteemed by ancient and modern Chinese societies; they symbolize love, affection, and feminine beauty, as well as having a notable position in Chinese pharmacopoeias. This species is known as the *zi ban mudan* (the purple-blotched tree peonies) in China.

The specific epithet acknowledges Joseph Rock (1884–1962), the Austro-American explorer, geographer, plant hunter, and linguist. The introduction of this plant to the West was fittingly romantic, with a figure no less passionate than Reginald Farrer first describing its charms as "the huge expanded goblets of *Paeonia Moutan*, refulgent as pure snow and fragrant as heavenly Roses" (Farrer 1917). Farrer saw this plant in southern Gansu in 1914, but Rock introduced it into Western cultivation in 1926. His discovery of the seed plant in the Choni (Jone) Lamasery in southwest Gansu led to seed being sent to the Arnold Arboretum in Jamaica Plain, Massachusetts, with seedlings eventually being sent to growers in the United States and Europe.

Our plants appear to be the typical American form of *Paeonia rockii* originating from Reath's Nursery in Michigan. The distinct English form, originating from Highdown Garden in Kent, is another seedling raised from the original Rock introduction.

This imposing shrub can grow into a huge sprawling specimen 2.4 m tall by 3 m across, with a distinctive bronzing to the young leaves before they assume a delightful bluish cast. The foliage contrasts superbly with up to 50 blooms on mature shrubs. The leaves are 2- or 3-pinnate with 17 to 33 leaflets. The leaflets are variable, ranging from lanceolate or ovate-lanceolate, entire leaves to mostly ovate-orbicular and lobed leaves (2–11 cm long by 1.5–4.5 cm wide). The leaf is glabrous on the upper surface and slightly tomentose on the underside, sometimes locally dense along the leaf veins.

The solitary, captivating flowers (13–19 cm) are borne terminally with 3 leaf-like bracts and sepals. The flower consists of 10 white petals, showing conspicuous dark purple blotches at the base. The filaments and flower disc are pale yellow. Flowering usually occurs in May. The fruits are delightfully reminiscent of a jester's hat, consisting of 5 oblong, densely yellow, tomentose, spindle-like follicles that split to reveal two rows of jet-black seeds.

The Qin Ling is the core area for this king of flowers—from the Loess Plateau area of Shaanxi–Gansu border, to the northern slopes of the sacred Taibai Shan, and finally to the diverse forests of Hubei's famous Shennongjia Forest District. This species grows in a wide range of ecological conditions, including sunlit openings in deciduous broad-leaf forests, margins of *Pinus armandi* forest and dense scrub, shady north-facing slopes, and limestone crags (Wang 1961). Forms with red and pink petals occur throughout the shrub's native regions, so there are rich opportunities for further flower selection.

HARDINESS: Zones 5–8. This species is hardy in maritime areas of the Pacific Northwest. Also, the natural range of this species occurs in areas with a markedly continental climate, perhaps indicating that the shrub could be grown more extensively in parts of the southern interior of British Columbia and south into the United States, where reasonable moisture is present. It appears to thrive best in the northern parts of the United States as far as Michigan.

CULTIVATION: The main challenge in the Pacific Northwest is excessive moisture in late winter and spring, which can promote attacks of peony blight (*Botrytis paeoniae*). Siting is a vital factor for this plant. It requires southern exposure, full sun, good air circulation, and quick drainage. Group planting should be avoided to prevent proliferation of the blight. Any sanitary pruning to remove blight damage or infected foliage should be done in late winter. This plant combines elegantly with other dry-land flowering shrubs such as *Berberis wilsoniae*, *Hypericum forrestii*, and *Rosa elegantula* 'Persetosa'. Acknowledgment must be given to the exciting work of Will McLewin of Phedar Nursery in the United Kingdom and his Chinese collaborators, Chen Dezhong and Cheng Fanyun. They are responsible for introducing to the West a new generation of *Paeonia rockii* forms and hybrids that are now gracing our gardens.

PROPAGATION: The seed of this species displays double dormancy, thus requiring a warm and a cold stratification to promote germination. These seeds should be given a warm stratification at 18 to 23°C for 3 months, followed by a cold treatment at 5°C for 3 months. Grafting is the preferred commercial method of propagation, especially for named cultivars.

Persea bracteata (Lecomte) Kostermans (Yu 7835)
syns. *Machilus yunnanensis* Lecomte, *Persea yunnanensis* (Lecomte)
    Kostermams, *M. yunnanensis* var. *duclouxii* Lecomte
Lauraceae   Plates 115–116

DISTRIBUTION: China, in western Sichuan, and in central, western, and northwest Yunnan, at 1500 to 2000 m.

IN GREEK, *persia* is the name for *Cordia myxa* (Sudan teak), which this tree resembles. *Persea bracteata*, a handsome willow-leaf shrub, has developed into one of the most distinguished broad-leaf evergreens in the David C. Lam Asian Garden. Every April, we are delighted to see the young leaves take the form of silver swords. During the early monsoon period in late spring (April to May), the spectacular colouring of the unfurling leaves of evergreen oaks and laurels in Southeast Asia is still lamentably unknown in the West. To quote the renowned plant explorer and geographer Frank Kingdon Ward (1949), writing on this oversight: "[They are] distinguished for their striking leaves, sometimes with an almost phosphorescent blue gleam beneath." A partial explanation for the oversight is no doubt because so much past botanizing occurred in early spring and fall, thus missing this moment of glory. I have been fortunate to witness this kind of display in the wild, and once seen, you are hooked for life.

Seedlings of this species were obtained from Washington Park Arboretum in Seattle during early 1987 from under a parent raised from seed collected by Yu (Yu 7835) and labeled *Persea yunnanensis*. The original collector, Professor T. T. Yu (1908–1986), worked for the Institute of Botany, Beijing, Chinese Academy of Science, and made this collection in the community of Heilongtan in the early 1950s, close to the present-day Kunming Institute of Botany. Yu collected for two British- and Irish-backed expeditions during 1937 and 1940 in Sichuan and Yunnan, despite world hostilities. He was a great authority on the genus *Sorbus*, with the outstanding *S. yuana* named in his honor. *Persea bracteata* is a common and widespread large, arborescent shrub to small tree in central Yunnan and is frequently utilized for home construction and furniture. In the wild it seems to favour slopes with moist, fertile soils under montane, evergreen, broad-leaf forests.

The species's lissome, linear-lanceolate leaves (10–21 cm long by 2–4 cm wide) are distinguished by a matte, bluish green cast to the upper leaf surface. The leaf underside is startlingly glaucous with a slight bluish stain contrasting beautifully with the yellow midrib. The petiole remains light green. The yellowish green younger stems are slightly flattened and ribbed, a common trait of lauraceous plants. The terminal buds (6 mm) are beaked like a bird, consisting of 2 elongated but unequal bud scales covered in fine, silky hairs. The specific epithet *bracteata* refers to the floral bracts that are borne below the flower clusters, which fall away as the flowers open in early May. The clusters of 2 to 3 racemose inflorescences are 12–14 cm long, each with up to 6 to 8 primrose yellow flowers grouped at the apex on a rachis up to 9 cm long. In fine years, small (7 mm), globose, black, fleshy berries are produced in some abundance.

The steeply ascending branching pattern of this species forms a beautiful dome of fine-textured, bluish green, finger-shaped leaves that on an overcast day have a curious silvery blue glow. Having spent time in the field in China viewing a host of lauraceous plants, I have concluded that this species is without doubt one of the most desirable evergreens for general cultivation.

HARDINESS: Zones 7–8. This species has survived the same minimum temperatures as *Persea ichangensis* in the Asian Garden with no damage observed. We have several specimens growing in sheltered and open situations, showing no signs of leaf bronzing or bark split after severe cold weather. Our experience with the species in the Pacific Northwest indicates that it has perhaps slightly more cold hardiness than *P. ichangensis*.

CULTIVATION: This species appreciates the same general cultural conditions as *Persea ichangensis* and a forest environment. In our garden, the shrub has thrived in a variety of sites ranging from full sun to deep shade. From Vancouver to southern Oregon, west of the Cascade Mountain range, this foliage plant of rare quality would be a fitting companion to evergreen aristocrats such as the fragrant and heavy flowering *Magnolia maudiae* or the incomparable ulmo, *Eucryphia cordifolia*, from Chile.

PROPAGATION: Until the development of a reliable seed source and/or advances in micropropagation for evergreen Lauraceae, these plants will be a difficult group to propagate. Many other ornamental lauraceous genera, including *Cinnamomum*, *Cryptocarya*, and *Machilus*, have potential, particularly in our warming world. T-budding, tip grafting, or wedge (cleft) grafting are used on robust lauraceous understocks for best results.

Persea ichangensis (Rehder & E. H. Wilson) Kostermans
syn. *Machilus ichangensis* Rehder & E. H. Wilson
Lauraceae   Plate 117

DISTRIBUTION: China, in Gansu, Guangxi, Guizhou, Hubei, Hunan, Shaanxi, and Sichuan, at 500 to 1400 m.

A REMARKABLE, large, evergreen shrub, *Persea ichangensis* is closely related to the well-known avocado (*P. americana*), one of over 150 species common to tropical and warm-temperate regions of the Old and New Worlds. This species was first collected by Augustine Henry (1857–1930), and then later by E. H. Wilson in 1900 and 1907 in western Hubei Province. There, it can grow into a tree 25 m high over time. The specific epithet *ichangensis* refers to the city of Yichang (Ichang), Hubei, on the banks of the Yangtze River, where Henry worked for the Chinese government.

In 1972, as a young horticultural student, I visited Wakehurst Place, Royal Botanic Gardens Kew's annex in Sussex, England, for the first time. There, I viewed this handsome evergreen with deep admiration. The delightful poise of the delicate finger-like leaves and refined, upright growth habit was unforgettable. In 1987 Tony Schilling, the curator, generously sent me scion-wood of one of the two plants growing at Wakehurst Place (000–69 31481) derived from Wilson seed. At UBCBG, we successfully grafted one plant onto *Machilus thunbergii*, which grew into a small shrub that was planted in protected deep shade, beneath a large grand fir (*Abies grandis*), since hardiness was unknown at the time. The plant has grown into a healthy specimen, but top growth has been diminished by intense root competition. It appears hardier than anticipated, and I look forward to acquiring future documented wild seed collections.

The leaves are generally narrow oblong-lanceolate, narrowing to an acuminate tip with a cuneate or attenuate base, including a 1.5–3 cm petiole (15–25 cm long by 2–3.5 cm wide). The coriaceous, glabrous, just-washed appearance of the upper leaf surface contrasts with the lighter, glaucous green of the leaf underside. The flowers are most distinctive and impressive for a lauraceous plant: inflorescences arise from the axils of prominent deciduous red bracts at the base of young shoots that consist of a cluster of up to 6 narrow panicles, each 10 cm long. The flowers are described as being white in the wild, but in cultivation yellowish green flowers are normal. The flowers are well presented, held above the graceful whorls of leaves. The fruit is a black, 1-cm berry, which in some years can be produced in abundance. David Hunt (1960) gives a detailed botanical history of the plant in *Curtis's Botanical Magazine*.

The famous plantsman E. H. Wilson reported this shrub as being widespread in Hubei, ranging to elevations of nearly 1500 m and often accompanied by another handsome evergreen, *Lindera megaphylla*, at slightly lower elevations. The introduction of *L. megaphylla* from Taiwan in our garden has lately suffered considerable leaf damage by temperatures around −10°C. Hubei provenances may be hardier, as is the case of the Wilson specimen at Borde Hill, Sussex.

HARDINESS: Zones 7–9. The true level of this plant's hardiness is slowly emerging in the Pacific Northwest. Our plant has survived unscathed at −13.4°C in December 1990 when a young plant, so Zone 7 may be its limit here. The coastal areas of southern British Columbia and south to the fog belt of Northern California may be an ideal home for this plant.

CULTIVATION: This evergreen thrives as an understory plant beneath the dappled shade of our West Coast conifers in the moderately acidic, well-drained, glacial till soil typical of the region. It appreciates moisture and loose surface organic matter. The shrub is not only a large, elegant choice for middle-ground and background situations, but it is suitable for up-close viewing. It is a soft complement to the trunks of our massive Pacific Northwest conifers.

PROPAGATION: Until the development of a reliable seed source and/or advances in micropropagation for evergreen Lauraceae, the plants will be a difficult group to propagate. Many other ornamental lauraceous genera, including *Cinnamomum*, *Cryptocarya*, and *Machilus*, have potential, particularly in our warming world. T-budding, tip grafting, or wedge (cleft) grafting are used on robust lauraceous understocks for best results.

Phoebe sheareri (W. B. Hemsley) Gamble
syn. *Machilus sheareri* W. B. Hemsley
Lauraceae   Plates 118–120

DISTRIBUTION: China, in Anhui, Fujian, Guangdong, Hainan, Hubei, Jiangxi, and Zhejiang.

THE LAUREL FAMILY, one of the structural linchpins of the mixed mesophytic and evergreen broad-leaf forests of central and southern China, comprises 24 genera and over 430 species in China, many of which are fine forest trees or shrubs. *Phoebe sheareri* is just one species in this vast, green, treasure trove I have grown to admire so much in cultivation and the wild. The specific epithet *sheareri* recognizes George Campbell Shearer (1836–1892), a Scottish doctor who took up residence in the British Concession of Kiukiang (now Jiujiang), Jiangxi, on the Lower Yangtze River in the 1860s. His scientific training and curiosity led him to botanize some of the mountainous areas, such as Lushan, close to his place of work. The generic name is of rather obscure Greek origin, commemorating the Greek goddess Phoebe, who was known for her ability to answer questions.

During 1988, I came across this species while exploring the lower elevation forests of the Huang Shan, one of China's sacred mountains in southern Anhui Province. I had just spied the remarkable climbing Chinese osage orange (*Cudrania cochinchinensis*) scrambling high into the forest canopy, when my gaze lowered and caught sight of a fine, substantial, multistemmed shrub with distinctive, bluish, obovate leaves. On closer examination each leaf had a short, curved, dagger-like drip tip, and was coated on the underside with short, rusty brown, downy hairs. In the wild and in cultivation, this species can grow into a tree close to 15 m, but generally forms a multistemmed arborescent shrub, typical in cultivation. The lower branches have a graceful horizontal pose that contrasts dramatically with the fan of stiffly ascending codominant leaders.

The leaves, even on the same plant, are extremely variable in size, ranging from 12 to 26 cm long and 4 to 10 cm wide. These remarkably shaped leaves are obovate to elliptic-obovate, with a dramatic apex drawn out into a curved acuminate tip. The leaf base is tapered, narrow, and cuneate. The upper surface is leathery matte green with some rusty-to-pink colouration to the midvein. The leaf underside has extensive rusty brown pubescence along the midrib and to a lesser degree along some of the secondary veins. Most veins on younger secondary leaves appear yellowish, while the interveinal areas have a slight bluish cast. The rusty brown pubescence extends to the petiole, younger branches, and to the rachis of the inflorescence. The flowers appear in June to early July, are borne in panicles (7–5 cm in diameter), and are small, greenish yellow, and ornamentally insignificant. Flowers are followed by black, egg-shaped fruits 5–6 mm in diameter in late September.

In the Huang Shan this species generally forms part of the forest understory, among trees including such varied companions as *Castanea seguinii*, *Nyssa sinensis*, and *Tapiscia sinensis*.

The genus *Phoebe* generally inhabits low-elevation forests in China that are characterized by heat, humidity, and heavy monsoonal rains. Many of these species, including *P. sheareri*, may also have a niche in parts of southern and southeastern United States with an analogous climate. The genus *Phoebe* is well known in China for providing fragrant, high-quality timber. *Phoebe nanmu* (Oliver) Gamble and *P. zhennan* S. K. Lee & F. N. Wei are frequently used in the construction of Buddhist temples.

HARDINESS: Zones 7–9. The first year these young plants were set out in the David C. Lam Asian Garden in 1990, they survived a temperature of −13.4°C in December of that year and then a lesser freeze to −12.5°C in December 1996. Our plants suffered some leaf and shoot damage in the 1990 freeze, but have since been untouched. This species appears capable of thriving in protected forest gardens in some areas of the maritime Pacific Northwest.

CULTIVATION: Our young plants of this collection (NA 60723) were planted in a protected, shaded location to replicate their native haunts. They have developed into an attractive, small grove just over 4 m high, showing the same kind of form and ornamentation so admirable in the wild. Since being planted, the shrubs have become increasingly impressive in their vigour and distinctive, handsome foliage. Moist, free-draining soil rich in organic matter with high overhead shade provides optimal growing conditions for *Phoebe sheareri*.

PROPAGATION: Until the development of a reliable seed source and/or advances in micropropagation for evergreen Lauraceae, the plants will be a difficult group to propagate. Many other ornamental lauraceous genera including *Cinnamomum*, *Cryptocarya*, and *Machilus*, have potential, particularly in our warming world. T-budding, tip grafting, or wedge (cleft) grafting are used on robust lauraceous understocks for best results.

Pileostegia viburnoides Hooker f. & Thomson
Cap-flower vine
Hydrangeaceae   Plate 121

DISTRIBUTION: China, in Anhui, Fujian, Guangdong, Guangxi, Guizhou, Hainan, Hubei, Hunan, Jiangxi, Sichuan, Yunnan, and Zhejiang, at 600 to 2000 m, and in Taiwan and Japan.

THE GENUS name *Pileostegia* comes from the Latin *pileo*, meaning "cap," and *stegia*, meaing "covering." The species epithet *viburnoides* refers to the resemblance of the leaves to the leaves of the genus *Viburnum*. China is home to a wide range of climbing hydrangeas, including such genera as *Schizophragma* and *Decumaria*. This refined evergreen species has graced the walls of numerous mansions in western and southern Europe, since being introduced by E. H. Wilson from western

Sichuan in 1908. In the wild it starts life as a sprawling ground-covering shrub before seeking and scaling a neighbouring tree or rock face. It climbs with the aid of aerial rootlets that allow it to attain 25 m into forest trees in the manner of ivy (*Hedera helix*). In early fall it has a moment of glory when showy, creamy white panicles shine forth like pools of light in the misty forest gloom.

The leathery, entire, lanceolate to oblanceolate leaves (10–22 cm long by 2.5–6 cm wide) are adaxially glabrous and lustrous, with the underside also being generally glabrous, with distinctive pitting to the surface. The leaf base is cuneate, with an acute apex. The young, unfurling leaves of rapidly extending shoots are often a stunning, brilliant, bronzy red colour. The flowers are clustered in large, congested, terminal panicles (7–20 cm long by 15–25 cm wide), which open during mid September and into October. The individual flowers are small, with the stamens being the most conspicuous element. The dry, hydrangea-like inflorescence can still be ornamental during the winter months, especially after a major snowfall.

The species in the wild assumes an imposing size, given time and a benign climate to suit its evergreen nature. In the fall of 2000, I had my first field introduction, a gargantuan specimen climbing up a vertical 20-m cliff in the Dulong (Taron) Valley in extreme northwest Yunnan. A few miles from Kongdan, at 1700 m, the Dulong Jiang is constricted by a gorge providing a perfect moist, almost subtropical habitat for this imposing climber. The evergreen leaves, with their pleated, layered appearance, dripped with moisture and shone with a pewter light at the foot of this gloomy cliff, sharing the space with ferns, gesneriads, and begonias. The second time I came across the cap-flower vine was in the Wuyi Shan Scenic Area, just outside Wuyi Shan City in northwest Fujian Province, during the fall of 2003. This is Robert Fortune country, where the famous Bohea tea has its origins, and is the home of Fortune's intensely sweet-scented finding, *Rhododendron fortunei*. In this context, the vine was in full flower and grew downward from a lofty perch 18 m above, before forming a cascading curtain over a huge, convex outcrop of gritstone. The frothy, creamy white blooms further enhanced the grace of this superb evergreen.

Another desirable species from southeast China, *Pileostegia tomentella* has delectable, dense, reddish brown indumentum on the leaf underside and still awaits introduction to the West.

HARDINESS: Zones 7–9. The UBCBG plants came to us in 1981 from Hillier Nurseries of Winchester, England. All plants suffered some minor leaf drop, while one plant suffered major leaf damage and bark splitting during an early November 1985 freeze of −12.7°C. The plants have grown vigorously since this early setback and have coped with lower temperatures since then. This shrub is a first-rate climber for the maritime Pacific Northwest and the southeastern United States. In marginal areas some winter protection for young plants would be prudent.

CULTIVATION: Several specimens are now clothing tree snags and living trees up to 17 m high in the David C. Lam Asian Garden. In flower they add a glorious dimension to the verticality of the surrounding forest, especially during the fall. This vine is a fine climber for residential gardens, walls, or concrete structures, as well as pergolas, patios, and arbors. Eastern and northern aspects are ideal, if more light is available as the climber ascends.

PROPAGATION: Cuttings should be taken in late spring to early summer. Even as a young plant this species frequently produces rooted shoots near the root crown. Later, young shoots with adventitious roots are also useful for propagation.

Pittosporum truncatum E. Pritzel ex Diels (SABE 1616)

Pittosporaceae   Plate 122

DISTRIBUTION: China, in Gansu, Guizhou, Hubei, Hunan, and Yunnan, at 300 to 2600 m.

STUDENTS OF plant distribution are often surprised that there are over 40 species of *Pittosporum* native to China, in a genus of over 200 species chiefly from Australasia but ranging across Africa to East Asia. *Pittosporum* is well adapted to low light conditions and often grows as shrubby understory or as an emergent tree beneath a forest canopy. The name *Pittosporum* is derived from the Greek *pitta*, for "pitch," and *spora*, for "seed." The seed has a sticky, resinous coating, a property that assists dispersal by animals or birds.

    Pittosporum truncatum develops into a distinctive upright shrub or occasionally a small tree with a short bole, up to 5 m under optimal conditions. This plant inhabits dense thickets, forests, and waterside locations. In deep shade in the wild it tends to have an open, wandering habit. The leaves are clustered in pseudowhorls at the branch tips with sparsely leafed internodes. Each distinctive leaf is hard, dark green, leathery, obovate or rhombic, and 5–8 cm long by 2.5–3.5 cm wide. The acute apex is often lobed, before narrowing rapidly at the waist and then forming a cuneate base. The flowers are solitary or in umbels, and are located in the leaf axils of lateral shoots. The creamy white, slightly fragrant flowers appear in early April, borne on slender pedicels. The fruit is a pea-sized, 2-valved capsule that splits at maturity to reveal orange seeds within a sticky pulp.

    The UBCBG received seed of this species in 1980 from the first Sino-American Botanical Expedition to the Shennongjia Forest District of Hubei Province, China. The seed was collected at about 1000 m near the town of Houshanping.

A closely related species, *Pittosporum brevicalyx,* was collected in 1980 by Stephen Spongberg (SAS 26B) near the Golden Temple, Kunming, in Yunnan Province. A selection named 'Golden Temple', made by the Saratoga Horticultural Research Foundation, San Martin, California, is proving to be a fine ornamental for Zones 7 to 9 of North America. The compact habit, evergreen leaves, fragrant yellow flowers, plus strong constitution make the shrub a highly desirable ornamental.

HARDINESS: Zones 7–9. A number of *Pittosporum* species from China appear much hardier than their antipodean cousins. *Pittosporum truncatum* is an example: it survived temperatures to −13.4°C in 1990 completely unscathed. Wider use of this species and perhaps others from China, therefore, is indicated, especially in the eastern United States south of Washington, DC.

CULTIVATION: One of the original seedlings growing in the David C. Lam Asian Garden is now a substantial, upright shrub of over 5 m growing in deep shade. The shade-loving qualities of Chinese pittosporums appear general in our experience, which makes these shrubs useful companions for the limited suite of woody shade lovers at our disposal. Observations in China indicate that a substantial number of species also appear to be well adapted to full sun. I have been impressed with the ability of *P. truncatum* to cope with dense shade and the rootwebs of large Pacific Northwest conifers. Well-drained, moderately acidic, glacial till–derived soil rich in surface organic matter seems to provide ideal conditions for this and other *Pittosporum* species.

PROPAGATION: Fresh seeds germinate readily with no pretreatment. Strong cuttings taken in early to mid July appear to give the best results.

Rhododendron asterochnoum Diels (EN 3551; C&H 7051)
subsection *Fortunea*
Ericaceae   Plates 123–124

DISTRIBUTION: China, in central Sichuan, at 1800 to 3660 m.

ORIGINALLY a name for the rose-flowered oleander, *Rhododendron* (from the Greek *rhodon* for "rose," and *dendron* for "tree") was later transferred by Linnaeus to this genus.

This shrub is just one of many fine species from China that have done particularly well in the Vancouver area since the reopening of China in the early 1980s. *Rhododendron calophytum* is closely related and is perhaps the most adaptable and hardy of the large-leaf rhododendrons in the Pacific Northwest. Although the botanical status of *R. asterochnoum,* a species introduced in 1990, is still open to speculation (Cox 1990), there is no doubt of its horticultural merit. *Rhododendron asterochnoum* has the ability to grow into a medium-sized tree of up to 15 m in the wild, while in cultivation, 10 m with optimal growing conditions is a more realistic expectation.

The specific epithet *asterochnoum* refers to the star-shaped, downy hairs that are a feature of the indumentum. This white to fawn, stellate indumentum extends to the young stems, mature petioles, and leaf undersides and is especially noticeable on young leaves. Although distinctive, it is often variable in density and colour, starting greyish white and maturing to fawn brown. In Edward Needham's Wolong collections, the whitish grey, stellate hairs at the leaf margins give a delicate, frosted appearance to the young foliage, which is especially prominent at the candle stage of development in the spring. The fine leaves are up to 15 cm long by 6.5 cm wide, oblong-lanceolate to oblong-elliptic. The inflorescence consists of 15 to 20 ivory-white to pinkish red flowers, with dark red blotches at the base. The margins of each flower are undulated and reflexed, giving them a dense, cauliflower-like appearance.

We are growing two wild collections of this interesting species in our garden. The first came to us in 1993 from Edward Needham from a collection (EN 3551) near the Wolong Panda Reserve, western Sichuan. The second came to us from a Peter Cox–Steve Hootman collection (C&H 7051) from the Daliang Shan of southern Sichuan. The vigour of our young plants from both collections suggests a mature plant size similar to *Rhododendron calophytum* with the same kind of site tolerances. The species has a distribution through mountainous areas of central Sichuan, often forming part of the understory beneath *Abies-Tsuga* forest.

Hardiness: Zones 6–8. This species has proved to be hardy in the maritime regions of the Pacific Northwest, with no damage yet recorded as of 2004. The mild winters of the last 10 years, however, have not given the plants a real test.

Cultivation: We have 20 individuals of this species growing well throughout the David C. Lam Asian Garden, revelling in the dappled shade of the high conifer overstory. In this setting it has proved a fine evergreen for many background locations; it associates well with other bold evergreens such as *Daphniphyllum himalaense* subsp. *macropodum* and foreground subjects like *Rhododendron degronianum* subsp. *yakushimanum* and *R. williamsianum*. Well-drained, moderately sandy, acidic soil with an abundance of surface organic matter are ideal for most large-leaf species. Organic mulch should extend at least 50 cm beyond the drip line of the shrub. Some level of high overhead shade is preferable, to provide both optimal levels of light for flower bud development and protection against moisture stress in late-summer drought conditions. Weevils, *Phytophthora cinnamomii*, *Armillaria ostoyae*, and the dreaded powdery mildew (*Chrysomyxis rhododendri*) are present in the Asian Garden. They pose a potential threat but have not afflicted our plants of this or any other rhododendron species discussed in this book to date.

PROPAGATION: Hand-pollinated or wild-collected seed is preferable, gathered October through January. Seeds should be sown in pans, which have been filled with moist peat with excess moisture allowed to drain off. The pans are then placed in a frame with bottom heat (18–21°C). With increasing light, germination should take place in 9 to 20 days. For detailed accounts of all propagation techniques for *Rhododendron*, see Peter Cox (1990, 307) and Michael Dirr and Charles Heuser (1987, 187). Thin cuttings from side growths are preferable, 7.6–10 cm long with 3 to 4 leaves retained. Cuttings should be taken in early morning when they are turgid, with larger retained leaves cut in half during mid July to early August. Rooting in peat-perlite gives satisfactory results.

Rhododendron denudatum Franchet (HM 1497)
subsection *Argyrophylla*
Ericaceae   Plates 125–126

DISTRIBUTION: China, in Central Sichuan, northeast Yunnan, and west and north Guizhou, at 1400 to 2700 m.

THIS RHODODENDRON is closely allied to three other first-class ornamental species, *Rhododendron coeloneuron*, *R. farinosum*, and *R. floribundum*, all of which appear to intergrade in the wild. All these plants adapt well to maritime parts of the Pacific Northwest where space allows. Collections of *R. denudatum* from western China introduced since 1993 are perhaps the pick of the bunch. The plants have impressed me by their vigour, outstanding foliage, and floriferousness. The specific epithet *denudatum* refers to the leaves, which become stripped of hairs on their upper surface as they mature.

In the wild this species can develop into a small tree of 8 m high or more, especially if drawn up by competing vegetation. The description given here is from a particularly fine collection by Bill McNamara of Quarryhill Botanical Garden, Lord Charles Howick, and members of a consortium including the Royal Botanic Gardens, Kew. Seed from HM 1497 was collected in October 1990 on Mount Luoji at 2860 m at the southern end of the Daliang Shan, Sichuan. The seed plants were huge arborescent shrubs, some up to 8 m, with some threatening to become trees. What determines whether some rhododendrons will become trees or shrubs in the wild is often the interplay of genotype, exposure, soil fertility, and age.

The leaves of our plants in cultivation from HM 1497 are generally 7–15 cm long by 3–5 cm wide, elliptic-lanceolate, bullate above, with a felted white or faintly rusty white indumentum, and leaf margins that are undulate and recurved. The spring combination of silvery white, spear-like, unfurled shoots contrasting with the mature foliage, long internodes, and a floriferous nature bodes well for the future of this plant in cultivation. The trusses, consisting of 5 to 12 individual flowers, are clear pink, with darker spots and blotches. In the wild, bluish lilac to pinkish purple colour forms of this species are also common. Another fine collection of this species by Peter Cox (C 5090), from near Kanding, west Sichuan, is also growing well in our garden. In the wild, *Rhododendron denudatum* inhabits dense, broad-leaf forests, up to near the tree line, where it forms windswept shrubberies at 2860 m in the Daliang Shan. Below that, it favours forest edges, glades, stream banks, or coppiced scrub.

The related *Rhododendron floribundum* differs in having narrower leaves, with a dullish matte patina on the upper leaf surface and paler indumentum. The older introductions of this species tended to have rather chlorotic foliage in cultivation.

Rhododendron coeloneuron is a glorious species that came to my attention in 1994 in the wilds of the Dashahe Cathaya Reserve in northern Guizhou. Peter Cox and others have also reported it from the nearby Jinfo Shan in southeast Sichuan. The upper leaf surface is glossy and bullate, with a dense, rufous, tomentose indumentum on the leaf underside. The textured leaves have a faintly sinister hooded look, which nonetheless is a fine foil for the clear pinkish purple flowers. Interestingly, this lovely species also occurs with *R. denudatum* in the Daliang Shan.

A final, fourth species less convincing to me as a separate species, *Rhododendron farinosum* is characterized by markedly hooded leaves, white flowers, and a more southern distribution in southern Guizhou.

HARDINESS: Zones 6–8. *Rhododendron denudatum* has proved hardy in the maritime regions of the Pacific Northwest, with no damage yet recorded in the five years it has been grown here. Ten consecutive years of mild winters, however, have not given the plants a real test.

CULTIVATION: In the David C. Lam Asian Garden, this species is proving to be a wonderful addition to our repertoire of tough, resilient, large-leaf rhododendrons that have year-round appeal. The altitudinal range, site tolerances, and morphological variation within this species and the three other mentioned related species makes them a promising foursome for use in our larger gardens. I particularly like the way they flower at an early age. They like to grow in intimate groups, and make excellent screens. We have a fine colony of *Rhododendron denudatum* growing around a giant cedar stump and set off by the aquamarine foliage of *Sorbus hupehensis*.

PROPAGATION: Hand-pollinated or wild-collected seed is preferable, gathered between October and January. Seeds should be sown in pans, which have been filled with moist peat with excess moisture allowed to drain off. The pans are then placed in a frame with bottom heat (18–21°C). With increasing light, germination should take place in 9 to 20 days. For detailed accounts of all propagation techniques for *Rhododendron*, see Peter Cox (1990, 307) and Michael Dirr and Charles Heuser (1987, 187). Thin cuttings from side growths are preferable, 7.6–10 cm long with 3 to 4 leaves retained. Cuttings should be taken in the early morning when they are turgid, with larger retained leaves cut in half during mid July to early August. Rooting in peat-perlite gives satisfactory results.

Rhododendron flinckii Davidian (KR 1442)
subsection *Lanata*
Ericaceae   Plates 127–128

DISTRIBUTION: Eastcentral and northeast Bhutan, at 3000 to 4000 m.

COMMEMORATING Swedish botanist Karl Evert Flinck, a leading expert on Rosaceae and especially the genus *Cotoneaster*, *Rhododendron flinckii* has proven to be very adaptable in cultivation and a fine foliage plant. It belongs to a group of species that appears to merge with several closely related species: *R. lanatum* from Sikkim and Bhutan and *R. tsariense* from Bhutan and southern Xizang (Tibet). For a review of this interesting group of species, see Cox (1990, 216, 358–360). The low-growing *R. tsariense* is perhaps the better known, and celebrated for its fine foliage and compact form though often proving fickle in cultivation.

The UBCBG specimens are derived from seed collected by Keith Rushforth (KR 1442) in 1988, from Bhutan, on the Rudong La (Pass) at 3992 m. These plants are happily established and have grown to 2 m. The leaves are beautifully formed, narrow elliptic, with a short acuminate apex. The upper leaves are darkish green with greyish brown indumentum that persists for a few months before being shed. The indumentum on the leaf underside is a startling orangey brown that glows, especially when struck by the low winter sun. The petioles and young stems are also covered in dense, light brown, tomentose hairs. For foliage alone, it has few peers and should be located where it can be appreciated. The flowers are creamy yellow with crimson spots.

HARDINESS: Zones 7–8. This species can be grown in Zone 8 if a favoured cool site is chosen, perhaps close to the sea. At higher elevations in the wild, burial by winter snow gives this plant additional protection. Our plants have endured −10°C without damage and I expect they could take −15°C or more.

CULTIVATION: This species grows naturally in a rugged alpine climate, often at the edges of fir forest among rocky outcroppings and boggy terrain. In cultivation it appears adaptable to sites from open to semishaded forest conditions. This species enjoys a cool, moist root run, so siting it in open situations would require some degree of natural moisture and shading of the surface roots. This species is a delight, especially for those who are fortunate enough to have moist, high-elevation, coastal gardens. It is a natural companion for many Pacific Northwest ericaceous plants, such as blueberries (*Vaccinium* spp.), that like cool, moist, root runs.

PROPAGATION: Hand-pollinated or wild-collected seed is preferable, gathered between October and January. Seeds should be sown in pans, which have been filled with moist peat with excess moisture allowed to drain off. The pans are then placed in a frame with bottom heat (18–21°C). With increasing light, germination should take place in 9 to 20 days. For detailed accounts of all propagation techniques for *Rhododendron*, see Peter Cox (1990, 307) and Michael Dirr and Charles Heuser (1987, 187). Thin cuttings from side growths are preferable, 7.6–10 cm long with 3 to 4 leaves retained. Cuttings should be taken in the early morning when they are turgid, with larger retained leaves cut in half during mid July to early August. Rooting in peat-perlite gives satisfactory results.

Rhododendron kesangiae D. G. Long & Rushforth
subsection *Grandia*
Ericaceae   Plate 129

DISTRIBUTION: India, in Arunachal Pradesh, and in Bhutan, at 2900 to 3500 m.

THIS SPECIES is without doubt the most impressive large-leaf species to be introduced to the Pacific Northwest since the 1980s. The quality of foliage, flower, and form is complemented by this species's adaptability to a broad range of site conditions. Curiously, the species was overlooked by earlier plant explorers, such as Ludlow and Sherriff in the 1920s and 1930s, who made no mention of its abundance and recorded it as a hybrid of *Rhododendron hodgsonii* and *R. falconeri*. Simon Bowes-Lyon introduced this species to the United Kingdom as seedlings given to him by the Kesang, Queen Mother of Bhutan, for whom it is named, in 1967. Keith Rushforth recognized the shrub as a separate species during his 1988 travels through Bhutan, and reintroduced it to Western cultivation. Subsequent exploration by Peter Cox and Steve Hootman has noted extension of the natural range of this species into neighbouring India.

Rhododendron kesangiae has the capacity to develop into a very large shrub, and with time and a humid climate could become a moss-shrouded, single- or multiple-trunked tree of about 6–8 m (12 m in the wild). I would expect these dimensions to be a real possibility, given 70 years in sheltered, coastal, rain-forest gardens of northwest coastal Washington State and the southwest coast of Vancouver Island.

The species has huge, broadly elliptic leaves (20–30 cm long by 10–16 cm wide), each rounded at the apex, with distinctive lateral veins on the upper leaf surface. The indumentum on the leaf underside varies from fawn to silvery when young, then fades with age. The young expanding shoots develop into brilliant silver spears, which must brighten the somber pre-monsoon forests of Bhutan. The terminal vegetative buds are most distinctive—rounded, jade green to dark red. The spongy-textured petioles are prone to wind breakage and so this species does best in sheltered forest gardens. The flowers are borne in dense trusses, consisting of 15 to 25 campanulate flowers, generally rose to pink, fading to light pink, with a large purple blotch and nectar pouches. Rich purple and white forms occur in the primeval *Abies-Tsuga* forests of its native range, often combining with the white-flowering crowns of *Magnolia campbellii*. In the last two years a few of our plants have started flowering, with deep pinks predominating.

HARDINESS: Zones 7–8. This species appears to thrive in the maritime areas of the Pacific Northwest, particularly in cool, hyperhumid, protected coastal areas, and south to the summer fog belt of northern California. Plants exposed to −10°C (in 2003) sustained no damage.

CULTIVATION: We have been growing this species since the early 1990s from numerous collections of Keith Rushforth and of Cox, Hutchinson, and MacDonald. The plants have proven to be outstanding in terms of ornamentation, adaptability, and hardiness. We have also found that this species grows most vigorously on water-receiving sites with fairly stagnant drainage, which mirrors closely the field conditions of its native land. In the David C. Lam Asian Garden, it is interesting to observe the decrease in internode length and leaf size of individuals growing in low, moist areas compared to the crest of knolls. This species would enjoy being planted with other large-leaf rhododendrons like *Rhododendron falconeri* and *R. hodgsonii*, which are associates in the wild. Interestingly, *R. kesangiae* inhabits the terrain above *R. falconeri* but below *R. hodgsonii*, with both species hybridizing with *R. kesangiae*. An overstory of massive Pacific Northwest conifers and ferns, and of bamboos is a great companion for this aristocrat of rhododendrons. The species appreciates heavy leaf litter.

PROPAGATION: If you cannot procure wild or controlled, hand-pollinated seed, grafting is the only alternative. Grafting can be done year-round, but February is the optimal month. In the case of large-leaf species, use healthy, strong understocks in the same subsection (*Rhododendron praestans* or *R. macabeanum*). For a detailed account of current grafting techniques, see Cox (1990, 315–316).

Rhododendron oligocarpum W. P. Fang (Guiz. 121)

subsection *Maculifera*

Ericaceae   Plate 130

DISTRIBUTION: China, in Guangxi and Guizhou, at 1800 to 2500 m.

ALTHOUGH GIVEN a slightly negative name, *oligocarpum*, meaning "with few fruits," this species is most attractive. *Rhododendron oligocarpum* was introduced into Western cultivation from a collection made on the Fanjing Shan of Guizhou Province in 1985. It has developed into a fine, symmetrical, robust shrub with many ornamental qualities and is very adaptable. I confirmed these observations in the wild, on a 2002 visit to the Miao'er Shan in northern Guangxi Province. There it forms rounded specimens crouching among granite boulders on a wind-blasted summit at 2100 m with the equally desirable *R. platypodum* as a companion.

In sheltered woodland gardens, this species could attain 5–7 m in height with age, while specimens grown in the open may only attain half that size. The leaves are oblong-oval, elliptic (6–8 cm long by 2–3 cm wide), with a mucronate apex, ciliate leaf margins, waxy leaf underside, and attractive reticulate venation. On the young leaves, the upper surface, leaf margins, petioles, and main vein are initially covered in buff hairs that are cast off later in the summer. The long internodes and clustered arrangement of the leaves toward the terminal buds are characteristic of this species in our forgiving climate. The leaves remain on this plant for up to 3 years, a sign of health and adaptability. The flowers of our specimen form small trusses (10 cm across) of 4 to 6 flowers, often borne on shoots that are partially hidden by new growth. The 5-lobed, campanulate flowers are light purplish mauve with prominent, dark purple basal blotches. Our plants have become increasingly floriferous over the past few years. The young candles of new growth are an elegant counterpoint to the lightly submerged flower masses.

In 1985, the Royal Botanic Gardens, Kew, organized an expedition consisting of Jim Russell, John Simons, and Hans Fliegner to explore the remote and mountainous Fanjing Shan, now a UNESCO/MAB World Biosphere Reserve, in northeast Guizhou Province. Their collections included several interesting rhododendrons, one of which, under Guiz. 121, was initially assigned to *Rhododendron maculiferum* subsp. *maculiferum* but has since been transferred to *R. oligocarpum*. This collection was from an exposed south-facing mountainside at 2100 m, growing in the open and making wide-spreading shrubs to 2 m. The form and foliage was reported as being similar to *R. williamsianum*. Jim Russell, the late curator and genius behind the Castle Howard Arboretum, Yorkshire, England, kindly sent me seed, which has resulted in our plants.

HARDINESS: Zones 6–8. This species appears to be another hardy one for the maritime Pacific Northwest. Our plant has shown no signs of distress from frost since it was placed in the David C. Lam Asian Garden in 1989. We do not normally suffer from late spring frosts that are the bane of many rhododendron-growing regions of the United Kingdom. This species like so many has colourful foliage displays, as the new candle-like shoots expand during spring growth.

CULTIVATION: This species appears to favour a more open situation than we have provided in the Asian Garden. Nevertheless it has shown itself to be a pest- and disease-free species that should be grown more widely for its distinctive foliage and charming pink flowers. Some of my favorite species, such as *R. maculiferum* subsp. *anhweiense* and *R. williamsianum*, have similar growth habits and integrate well with this attractive newcomer. Larches (*Larix* spp.) with their light leaf fall associate well with these species, providing dappled shade and an aesthetic counterpoint.

PROPAGATION: Hand-pollinated or wild-collected seed is preferable, gathered betwen October and January. Seeds should be sown in pans, which have been filled with moist peat with excess moisture allowed to drain off. The pans are then placed in a frame with bottom heat (18–21°C). With increasing light, germination should take place in 9 to 20 days. If seed is not available, cuttings are the best option. Thin cuttings from side growths are preferable, 7.6–10 cm long with 3 to 4 leaves retained. Cuttings should be taken in the early morning when they are turgid, with larger retained leaves cut in half during mid July to early August. Rooting in peatperlite gives satisfactory results. For detailed accounts of all propagation techniques for *Rhododendron*, see Peter Cox (1990, 307) and Michael Dirr and Charles Heuser (1987, 187).

Rhododendron sinofalconeri I. B. Balfour (SEH 229)
subsection *Falconera*
Ericaceae   Plate 131

DISTRIBUTION: China, in southwest Yunnan, and also in adjacent Vietnam, at 1600 to 3000 m.

THE DIRECT translation of the species name refers to this plant's close relationship with the Himalayan *Rhododendron falconeri*. Although introduced from northern Vietnam into Western cultivation by Keith Rushforth in 1992, only since 1998 has this species been planted in the collections of rhododendron connoisseurs. It is one of two introduced large-leaf species showing great promise at UBCBG. *Rhododendron sinofalconeri* displays fine foliage with striking yellow flowers, possibly matching the renowned, related *R. macabeanum*. This species holds great promise for hybridizers. We are currently growing collections made by Peter Cox and Steve Hootman under SEH 229 in 1995, collected on the Laojun Shan, near Wenshan, at 2900 m in southeast Yunnan close to the Vietnam border. Our young plants are thriving and appear to appreciate our temperate coastal climate.

This species can attain greater size than *Rhododendron falconeri*, forming a large, compact, arborescent shrub, and it can ultimately grow into a tree 15–20 m high under favourable native conditions. In cultivation, its maximum height is likely to be 7–9 m if shelter, rainfall, and humidity are ideal over a period of 70 years. Many of these large-leaf species can live for hundreds of years under favourable conditions. Frank Kingdon Ward mentions in his book *Return to the Irrawaddy* (1956, 166) that large *R. arizelum* can live perhaps 300 years, which would also apply to other megarhododendrons like *R. protistum* and *R. sinogrande*.

The leaves of this handsome species are broadly obovate, tending to become elliptic at maturity, and up to 28 cm long. The leaf underside is covered in a light brown, woolly indumentum, while the upper surface is glabrous, wrinkled, or rugulose. The flowers are arranged in a 10- to 12-flowered truss, with each campanulate flower coloured pale to rich yellow. *Rhododendron sinofalconeri* differs from *R. falconeri* by having a more obovate leaf and paler indumentum.

In the wilds of southeast Yunnan and in the Fan Si Pan range of neighbouring Vietnam, this species grows along open, cool, misty, knife-edge ridges where it can form pure elfin forests of up to 8 m, at an elevation of 2500 to 3000 m (Hootman 2003). Descending from this zone, *Rhododendron sinofalconeri* fights to become a large forest emergent, competing with evergreen oaks, magnolias, and maples.

HARDINESS: Zones 7–8. This newer introduction is already showing signs of being much hardier than would be expected of a plant from such a southern latitude. Martyn Dickson has documented that at the Royal Botanic Garden, Edinburgh, collections from Wenshan in southern Yunnan have survived to −12°C with no damage. A cold snap in Vancouver in December 2003 confirms that level of hardiness. This finding is most encouraging and indicates adaptability to the maritime, cool summer nighttime temperatures of the Pacific Northwest and perhaps south to the summer fog belt of northern California.

CULTIVATION: This shrub is a first-rate large-leaf species for moist, sheltered coastal gardens in the Pacific Northwest. In the David C. Lam Asian Garden, it thrives and associates well with other broad-leaf evergreens and smaller leaf, shade-tolerant trees, such as *Acer caudatifolium* and *Sorbus caloneura*. It appreciates moist but well-drained, humus-rich soils, ideally among moss-encrusted stumps and decaying logs. Like so many large-leaf rhododendrons, it is also in its element overhanging fern-banked torrents, waterfalls, or plunge-pools.

PROPAGATION: If wild or closed pollinated seed cannot be procured, grafting is the only alternative. Grafting can be done year-round, but February is the optimal month. In the case of large-leaf species, use healthy, strong understocks in the same subsection (*R. falconeri* or *R. hodgsonii*). For a detailed account of current grafting techniques, see Cox (1990, 315–316).

Ribes davidii Franchet (DJHC 777)
David gooseberry
Grossulariaceae   Plates 132–134

DISTRIBUTION: China, in northern Guizhou, southwest Hubei, southwest Hunan, Sichuan, and northwest Yunnan, at 900 to 2700 m.

THE GENERIC name *Ribes* comes to us from the Arabic or Persian word *ribas*, meaning "acid tasting," which is descriptive of cultivated gooseberry. This diminutive gooseberry belongs to a small, specialized group of about five species from central and western China that are evergreen, generally epiphytic, and thrive in cool cloud forests that abound in that region. The most familiar species of this group in Western cultivation is *R. laurifolium*, which in milder areas of Western Europe can form a fine, evergreen wall shrub with yellow-green flowers in racemes that often retire beneath the leaves. Steve Hootman's collection from southern Sichuan, under SEH 109, belongs to this species. *Ribes davidii* is named after the French missionary and plant explorer, Jean Pierre Armand David (1826–1900), whose name graces numerous choice Chinese plants. Dan Hinkley collected this dainty forest-dweller on the summit of Emei Shan among mossy rocks and tree trunks.

This species has the polymorphic ability to assume the form of a terrestrial shrub, ground cover, bark scrambler, or epiphyte in moist, forgiving, cloud-forest environments. This evergreen species, depending on what substrate it is growing on, ranges from a ground cover to a 1-m-tall shrub. The branchlets are spreading in habit, devoid of prickles and hairs, and terminate in prominent ovoid buds. The leaves (2 to 5) are clustered at the branch tips, with each leaf folded at the margins. The handsome leaves (2–3 cm long by 1.5–3 cm wide) are leathery and hairless, generally obovate-elliptic, with a glossy sheen. They are prominently 3-veined, with a narrowly cuneate base and blunt apex. The leaf margins are coarsely crenate-dentate toward the apex.

The male flowering racemes are erect, while the female flowers are axillary. The greenish white to yellowish green flowers in racemes that can be up to 7 cm long, appear in April to May. Plants in the David C. Lam Asian Garden have not yet set the ellipsoidal, purple fruits of this species.

In June 1993, Roy Lancaster saw *Ribes davidii* in the wilds of Hongya County, southwestern Sichuan, growing as an epiphyte in cloud-forest communities at 2700 m, with *Rhododendron moupinense*, *Vaccinium moupinense*, and the intriguing pink-flowered, epiphytic Solomon's seal, *Heteropolygonatum xui*. The David gooseberry grew there on the upper branches of *Abies fabri*, some 19 m from the ground, and was only identified by the discovery of a storm-tossed branch. The yellow, tubular flowers contrasted with the broad, fleshy leaves that were purplish on the underside. It is clear there is considerable variation throughout its natural range that includes diverse habitats. Dan Hinkley's fall 1996 collection (DJHC 777; see Plates 133 and 134) was only spied and then collected because some of the older leaves had turned a brilliant autumnal crimson and shone forth in the gloom. It grew on the mossy forest floor with *Hydrangea anomala*, and both were striving to hitch a ride up a neighbouring fir. The individual flowers of this collection have an unusual colour for this species. Each flower consists of an elongated, yellowish green corolla tube, crowned by a vermilion red lip. The underside of the fleshy leaves is light yellowish green.

HARDINESS: Zone 6–8. The altitudinal range of this plant is considerable, so some variation exists. This species is a temperate, cloud-forest shrub, which suggests adaptability to the cool coastal areas of the Pacific Northwest and south to perhaps Northern California's redwood forests. High summer nighttime temperatures in eastern and southeastern United States could be problematic for this species.

CULTIVATION: We are growing this plant successfully as a ground cover in shade with Japanese spurge (*Pachysandra terminalis*) and *Polystichum munitum*. Welldecayed stumps, logs, or even live trees in hyperhumid environments are planting opportunities for this pliable species. In summer-dry temperate regions, gardeners should take advantage of naturally moist but well-drained sites.

PROPAGATION: Seeds maintain their viability for a long time. A pretreatment of cold stratification is recommended for good germination. Softwood cuttings taken in July root well if misted and grown in a sandy medium.

Rosa multibracteata W. B. Hemsley & E. H. Wilson (SICH 096)
Rosaceae   Plate 135

DISTRIBUTION: China, in Sichuan and Yunnan, at 2100 to 2500 m.

THIS FINE upright rose quickly attains 3 m or more in 10 years. The plant is notable for its attractive upswept branches that are covered with an armature of paired spines. The leaflets are numerous (5 to 9) and small (1.5 cm), generally rounded and toothed, creating a delicate ferny appearance. This foliage is a perfect foil for the numerous flowers, which are generally borne in terminal corymbs or panicles. Each flower is 3.75 cm across and bright shell pink in colour, and is subtended by conspicuous leaf-like bracts whose presence is acknowledged in the specific epithet. The flowers are followed by fine, orange-red, bottle-shaped hips (6–10 cm).

This species has an extensive range throughout western China, notably western Sichuan, in the Min River drainage at 1600 to 3000 m, often favouring forest margins. In 1908, E. H. Wilson introduced this species to Western cultivation from northwest Sichuan. The plants in our garden originated from collections made near Maowen, Sichuan, under SICH 096, collected in 1988 by a consortium that included the Royal Botanical Gardens, Kew; Quarryhill Botanical Garden, California; and Howick Arboretum, Northumberland.

In the Pacific Northwest, notable features of the species include its sequential flowering from May to early August, the quantity of flowers borne at any one time, and its graceful cascading appearance. The plant is a wonderful choice for well-drained sandy soils with full exposure. *Rosa multibracteata* is closely related to *R. forrestiana* Boulenger, with many of its ornamental qualities including fine bright pink flowers and an extended flowering period from June to July. That shrub is another fine introduction made by George Forrest with a natural distribution in southwest China.

HARDINESS: Zones 6–9. This tough plant ranges close to the Tibetan Plateau and is accustomed to drought and the brunt of frequent cold waves pouring off the surrounding high ground. It would not be suited to the high humidity and heat of eastern North America. It is quite hardy in the Pacific Northwest, even in some areas east of the Cascade Mountains.

CULTIVATION: This species is adapted to disturbed, open sites with quick-draining soils. Like many shrubby rose species, it is a useful plant along disturbed southern and western fringes of our native evergreen forests where drought, direct sun, and intense root competition provide an ideal niche. Specimens at the Asian Garden were introduced in 1980 from Hillier Nurseries, Winchester, England. This plant has impressed me by its vigour and resistance to pests and diseases. Great companion species (all requiring dry sites) include *Hypericum forrestii*, *Osteomeles schwerinae*, *Rostrinucula dependens*, and *Syringa laciniata*.

PROPAGATION: Seeds should be collected when the hips are red. Seed coats are hard and often require acid scarification, followed by a 3-month cold treatment for adequate germination. Softwood cuttings root readily in a peat-perlite mix with mist.

Rosa willmottiae W. B. Hemsley
Helen Willmott rose
Rosaceae   Plate 136

DISTRIBUTION: China, in Gansu, Qinghai, Shaanxi, and Sichuan, at 1300 to 3800 m.

THIS VIGOROUS rose has the ability to become a massive upright bush of 4 m high and across. As it approaches maturity, it can scramble and arch into the lower branches of trees. The young, strong, ascending then arching stems are initially glaucous with pairs of straight prickles, but then turn dark reddish brown with age. The handsome bluish green leaves are composed of 9 leaflets and act as an attractive counterpoint to the profusion of flowers that are produced sequentially from early June to mid August. The dainty flowers are small (2.5–3.75 cm), clear rose, bright rich pink to rose-purple, with white stamens, and are borne on short lateral shoots. The bright orange-red oblong hips are often produced in compelling cascading masses. The species is named after a nineteenth-century English rosarian who supported E. H. Wilson's expeditions to China.

The species is abundant in northern Sichuan especially along the imposing flanks of the Min Shan range on the border with Gansu province—Reginald Farrer country. Populations occur just above the town of Songpan on the Min River, which was known as Sungpan when E. H. Wilson used it as a base for his explorations in 1913. In early June the flowering species sprawls into surrounding scrubby entanglements, creating a painter's dream by streaking the hillsides pink. A recent visitor to the region mentioned that this species grew in the company of *Paeonia veitchii* with pale pink flowers, the pink-flowered Holy Grail of alpine plant enthusiasts *Stellaria chamaejasme*, and the rather sinister green and black flowers of the solanaceous *Scopolia tangutica*. In this vast rugged land, it is also a likely associate of two other stunning ornamentals, *Dipelta elegans* and *Paeonia rockii*. A fine new introduction (SICH 208) has been made from near Jiuzhaigo, Sichuan, by a consortium that included the Royal Botanical Gardens, Kew; Quarryhill Botanical Garden, California; and Howick Arboretum, Northumberland.

HARDINESS: Zones 6–9. This shrub is another tough resilient species from northwest China that can cope with severe winter conditions. It is perfectly hardy for the Pacific Northwest, with similar general tolerances as described for *Rosa multibracteata*.

CULTIVATION: In the wild, like so many of its ilk, it is at home in thickets, scrub, rocky terrain, and streamsides. This rose shows considerable adaptability to both open and semishade conditions since its introduction to the UBCBG in 1980. It has shown the ability to grow in dense shrubberies and to clamber into the lower branches of large trees. One delightful example at UBCBG is a huge specimen now sprawling into a large *Acer griseum*.

PROPAGATION: Seeds should be collected when the hips are red. Seed coats are hard and often require acid scarification, followed by a 3-month cold treatment for adequate germination. Softwood cuttings root readily in a peat-perlite mix with mist.

Rostrinucula dependens (Rehder) Kudō (DJHC 664b)

Lamiaceae   Plates 137–138

DISTRIBUTION: China, in Guizhou, Shaanxi, Sichuan, and Yunnan.

THE GENUS name *Rostrinucula* roughly translates "with a small beak," and the specific epithet *dependens* refers to "being suspended" or "hanging down." I first learned of this plant when reading an account by Jim Russell of collections he made while participating in the 1985 Sino-British Expedition to the fabulously rich Fanjing Shan World Biosphere Reserve in northeast Guizhou Province. He kindly sent me a plant (Guiz. 018) from his collection, which unfortunately languished in a shady part of our garden for a few years and eventually died. I considered this plant a failure until we grew a collection made by Dan Hinkley, misidentified as a *Buddleja* species under his number DJHC 664b. This individual has proven to be a cracking good plant. The genus *Rostrinucula* is composed of two species, *R. dependens* and *R. sinensis*; both inhabit arid, scrub, hill country through parts of central China. Interestingly, Guiz. 018 now appears to be *R. sinensis*.

This splendid suckering shrub has grown to a symmetrical mound 2.5 m tall by 3 m wide, with the outer branchlets hanging closely downward at the fringe, emphasizing its growth habit. The smaller ornamental details are worth relating, as they contribute so much to the charm of this plant. The upper surface of the pendulous, wand-like outer branches is a startling reddish burgundy, which also stains the short petiole, the leaf midvein, and the extreme outer margin of the leaf. The leaves (4–7 cm long by 1–4 cm wide) are generally oblong-elliptic, papery, with a conspicuously bullate upper surface. They are reflexed with a short, twisted, acuminate apex; the larger mature leaves have a conspicuous cordate base. The vegetative buds and leaf underside are covered with white stellate-tomentose hairs, which offer a superb contrast to the burgundy stems. Fall colour is often late and an undistinguished yellow.

The flowers are produced in abundance during the late summer and consist of long-lasting pendulous verticillasters (6–27 cm long by 1.5 cm wide), the characteristic floral arrangement of the mint family. These delightful catkin-like inflorescences consist of beak-like, grey-white bracts densely covered with white indumentum. The bracts fall away as the protruding rose-purple flowers open sequentially toward the apex of the verticillaster. The young, developing, chalk-white "catkins" create a passing visual echo to the male catkins of *Garrya elliptica*.

Our DJH plant was collected near the Wolong Nature Reserve, the famous panda refuge, in Sichuan. This shrub grew in groups to 2.8 m in an open area close to the entrance of a gorge. This particular site was also memorable for the presence of *Epimedium davidii* and for golden monkeys (*Rhinopithecus roxellanae*) bounding through the trees.

HARDINESS: Zones 7b–9. This species is a superb ornamental, having survived −10°C with only minimal damage to branch tips in the winter of 2003 to 2004. Locations along the southern shores of the Strait of Georgia in British Columbia and south to Seattle should be ideal for this species.

CULTIVATION: The combination of a stunning cascading profile and highly ornamental leaves and flowers makes this shrub a good candidate for hot, sunny locations with sandy, quick-draining soils. I can see the plant as a natural companion to other shrubs of similar profile, such as *Buddleja alternifolia*, *Kolkwitzia amabilis*, and *Rosa elegantula* 'Persetosa', and perhaps fringed with Asian lilies, such as *Lilium duchartrei* or the sublime *L. regale*.

PROPAGATION: The suckering habit of this plant allows propagation from rooted offsets (ramets), which can be separated from the parent plant with a sharp spade and potted up. Cuttings taken in June and July and placed in peat-perlite with mist will root readily. The same prescriptions for the closely related *Elsholtzia stauntonii* have shown success.

Sorbaria tomentosa (Lindley) Rehder var. *tomentosa*
Lindley false spiraea
Rosacaeae   Plate 139

DISTRIBUTION: Afghanistan, central Nepal, central Asia, and northern China, at 1800 to 2900 m.

THE GENERIC name *Sorbaria* alludes to this plant's foliar resemblance to *Sorbus* (mountain ash), and the vernacular name acknowledges a passing similarity to the shrubby genus *Spiraea*. For those who appreciate a giant specimen in their gardens, this imposing shrub has muscle in leaf and flower. Our plant originated from Hillier Nurseries of Winchester, England, in 1979; it is correctly named *Sorbaria tomentosa* var. *tomentosa* (with leaf pubescence), as opposed to the Kashmir false spiraea (*S. tomentosa* var. *angustifolia*) which is glabrous. In the wild it grows across vast areas of central Asia, often favouring riparian sites, thus enabling it to range into the rain shadow of the Himalayas and other mountain massifs of the Tibetan Plateau.

The Lindley false spiraea naturally is gregarious, wide spreading, often suckering, and eventually forming a massive irregular dome of foliage to 6 m high. The stems are glabrous, with a waxy bloom over a finely ribbed surface. They arch upward and assume a more horizontal pose at flowering maturity. The young unfurling shoots in the spring become a bright bronze-red before they expand. The massive frondose leaves that follow are 20–40 cm long, consisting of a rachis with 11 to 21 opposite leaflets, giving this shrub a slightly tropical appearance. Each leaflet (11 cm long by 3.5 cm wide) is lanceolate to narrow-acuminate, with biserrate margins, light brown pubescence on the underside, and silky to the touch. The most interesting micro-ornamental feature is the beautifully serrated, elongated, filamentous leaf tip.

The huge, creamy white, plume-like flower panicles can measure up to 45 cm and are a fabulous spectacle. They appear in midsummer at the terminals of the current season's growth when other shrub plantings often start to look tired. The flowers slowly turn brown as they senesce, but remain impressive throughout the fall, and in winter they can appear to be resurrected after a heavy snowfall.

HARDINESS: Zones 5–7. This shrub is reliable for cold locations. It has application on both coasts of North America except the southeastern United States.

CULTIVATION: This resilient ornamental thug of a plant copes with difficult, saturated soils and tolerates air pollution, yet still has many refined ornamental attributes. Waterside habitats are ideal, and it looks happy with other hydrophiles like *Cornus alba* 'Sibirica', *Viburnum opulus*, or *Spiraea douglasii*, but my favorite companions include *Hydrangea aspera* Villosa Group and *Koelreuteria paniculata*. It is a first-rate subject for erosion control along riverbanks and could be a natural companion to *Filipendula kamtschatica*. To encourage heavy flowering, shorten old blossoming shoots and remove weaker suppressed branches in early spring.

PROPAGATION: Fresh seed will germinate without pretreatment. Generally this shrub is easy to root within 1 month from summer softwood cuttings placed in a peat-perlite mix with mist.

Sorbus aronioides Rehder (SICH 407)
Rosaceae   Plates 140–141

DISTRIBUTION: China, in Guangxi, Guizhou, western Sichuan, and western Yunnan, at 1000 to 3600 m, and in northern Myanmar (Burma).

THIS DELIGHTFUL, newly introduced species can become arborescent given time and favourable circumstances, but in general it makes an attractive flat-topped shrub. It is one of many *Sorbus* introductions from China and the Himalayas since the 1980s that have both enriched our gardens and intrigued taxonomists by their considerable charm and complex diversity. *Sorbus aronioides* belongs to the section *Aria*, which includes all the simple-leaf species, in contrast to the better known compound-leaf mountain ashes. The section *Aria* has even been occasionally treated as a distinct genus. Interestingly, the generic name *Sorbus* is derived from *sorbum*, the name for the fruit of the European service tree, *S. domestica*.

We raised a single plant from seed collected in Sichuan in 1988. Our plant has grown into a broad, wide-branched specimen of nearly 3.6 m tall by 4 m wide in a sunny clearing backed by other large shrubs. In the wild it appears versatile as far as site requirements are concerned, ranging from broad-leaf mixed forests to exposed mountainsides.

The leaves (10–15 cm long by 5–6 cm wide) are handsome and distinctive, resembling those of the genus *Aronia* (chokeberry), hence the specific epithet *aronioides*, meaning "aronia-like." The simple leaves are elliptic-obovate, with serrate margins, 7–8 prominent pairs of alternate leaf veins, a cuneate tapered leaf base, and a prominent twisted apex. The dark green leaf margins are in-rolled, creating a cleancut texture to the foliage. The purplish brown branchlets are conspicuously speckled with grey-white lenticels that diminish in size along the pedicels of the terminal compound corymbs.

The corymbs are dense and white flowered up to 7 cm across and appear in early May. The slightly pear-shaped, pea-sized fruits have an attractive one-sided purplish bloom. The other half of the fruit remains a yellowish green until final ripening. The fruit is borne in large clusters at the branch terminals, providing a strong contrast to the subtle yellowish orange fall colour.

The UBCBG plant is derived from a collection gleaned from wind-pruned, shrubby trees growing on an open mountainside on the Relong Shan (near Luding), Sichuan, at 2400 m; this collection, under SICH 407, was made by a consortium that included the Royal Botanical Gardens, Kew; Quarryhill Botanical Garden, California; and Howick Arboretum, Northumberland.

Another remarkable shrubby species that should be considered especially in the Pacific Northwest is *Sorbus insignis* (J. D. Hooker) Hedlund from the mountain forests of the Three Corners, where Yunnan, Myanmar (Burma), and Xizang (Tibet) meet. That species thrives as an epiphyte in the high crowns of forest giants or on mouldering stumps. Waxy, dark green, compound leaves and stunning, waxy, red fall colour make *S. insignis* a must for growing in mild areas.

HARDINESS: Zones 6–8. *Sorbus aronioides* will grow well in Pacific Northwest areas west of the Cascade Mountains from the subalpine to the seaside.

CULTIVATION: This species has proven adaptable, is free of disease, and is a fine ornamental. The key to growing most *Sorbus* species, including this one, is to provide a reasonably open site with a cool, moist, free-draining rooting environment that never dries out in late-summer drought. Too often these species are consigned to an early death by inappropriate site selection. This plant looks especially attractive in a moist, shrub-fringed dell surrounded with dramatic herbaceous companions including *Rodgersia*, *Hedychium*, and graceful colonies of *Polygonatum*.

PROPAGATION: From seed, 3 months of warm stratification followed by 3 months of cold stratification gives good results. Whip grafting in February should be considered using understocks such as *Sorbus aria* or *S. caloneura*. The understocks should be root- and stem-pruned before grafting. After grafting, the union should be waxed.

Stachyurus salicifolius Franchet
Willow-leaf stachyurus
Stachyuraceae   Plates 142–143

DISTRIBUTION: China, in Sichuan, Yunnan, Guizhou, and Guangdong, at 800 to 2000 m.

THE GENUS name *Stachyurus* comes from the Greek *stachys*, meaning "ear of corn" or "spike," and *oura*, meaning "tail," in reference to the form of the racemes. This introduction from China has quickly become a favourite of aspiring plant collectors because it combines grace with year-round aesthetic appeal. The late J. C. Raulston (1940–1996) of North Carolina State University introduced the plant from seed derived from Shanghai Botanical Garden in the early 1980s. This lovely, generally deciduous shrub is vigorous and quickly assumes the graceful weeping habit that sets it apart from other species. In the wild, *Stachyurus* often forms a portion of the understory shrubs that vie for space beneath the towering forest canopy in southern central China. The shrub can also cope with forest disturbance and is often a component of the scrub that results after logging.

The plant is capable of forming an irregular beehive-like form, growing to over 4.5 m high and perhaps a little wider. It catches the eye, displaying willowy pleated linear leaves (hence the epithet *salicifolius*) that hang vertically before ending in recurved tips. The leaves measure 8–20 cm long by 0.6–1.5 cm wide, are finely toothed, glabrous, glossy green above, and striped with a reddish pink midrib, which further enhances the willowy texture of the foliage. In some seasons, a semideciduous habit can be displayed, with only some leaves turning a respectable yellow in late November. The young twigs and flower rachis are also stained, providing further contrast to the leaf and flower colour. The yellowish cream flowers, consisting of 4 sepals and petals and contrasting green stamens and stigmas, are borne from the leaf axils in stiff, pendulous, vertical racemes. They consist of 12 to 25 flowers on racemes up to 8 cm long, distributed regularly along the branchlets and occasionally in terminal clusters. Flowers open from late February to mid March, but a late season can delay this by several weeks. The fruits are generally inconspicuous, consisting of green berry-like aggregates, each hanging down from the underside of the red-stained branches.

HARDINESS: Zones 7b–8. A cold spell to −10°C in December 2003 has confirmed some degree of frost hardiness. Our plant was deciduous at the time, and has shown no ill effects from the event. A number of related evergreen species are being trialled in the Pacific Northwest, notably at Heronswood Nursery in Washington State, so we should soon have a clearer idea of their hardiness status. This species requires a protected site away from cold desiccating winds and early morning winter sun, which can thaw frozen tissue too quickly, causing severe damage.

CULTIVATION: The appealing features of this plant are numerous, including a form that is graceful, with handsome foliage and flowers to match. This lovely woodlander should be planted in cool, moist, semishaded shrubberies with a background of conifers or broad-leaf evergreens. It is especially effective for counteracting the sometimes overpowering presence of big-leaf *Rhododendron* specimens, such as *R. falconeri* or *R. sinogrande*. An early season trio I look forward to planting includes this species with *Corylopsis pauciflora* and *Hamamelis mollis*. In smaller urban gardens, the shrub complements the grace and poise of *Camellia transnokoensis* or colonies of lilies such as *Lilium monodelphum*. Generally, *Stachyurus* is moderately lime tolerant, though it tends to prefer moist, acidic, humus-rich soil.

PROPAGATION: Propagation is best accomplished by taking semiripe-wood cuttings during July and using bottom heat.

Staphylea holocarpa var. *rosea* Rehder & E. H. Wilson (EN 3625)
Staphyleaceae   Plate 144

DISTRIBUTION: China, in Guizhou, Hubei, Sichuan, and Yunnan, at 1300 to 2500 m.

AUGUSTINE HENRY, the pioneering Irish physician and botanist, came upon this lovely arborescent shrub in the wilds of Hubei before the intrepid "Chinese" Wilson introduced it to the Arnold Arboretum in 1908. The genus *Staphylea* comes from the Greek *staphyle*, meaning "a cluster," and the epithet *holocarpa* means "whole fruited" (not lobed or split). The pink-flowered form of this species is a showstopper in the Pacific Northwest, benefiting from our sunny, warm summers and benign winters. The large arborescent shrub (megashrub) is a member of the bladdernut family, so called because of the inflated seed capsules. If grown as a tree, it has the appearance of a young specimen of the related endangered *Tapiscia sinensis*. Our plants are derived from seed collected on Emei Shan, Sichuan, by Edward Needham under EN 3625 in 1992. There at 1500 m Needham found this shrubby tree growing on a steep south-facing slope in open, mixed woodland containing many plant treasures including *Magnolia szechuanica* and *Styrax hemsleyana* within a matrix of maples, laurels, and rhododendrons.

In the wild, this species can reach up to 10 m, with sensuous smooth bark that is finely striped with chalk white serpentine vertical markings that shine dramatically after rain. The elegant trifoliate leaves appear clustered at the branch tips and are separated by wide internodes up to 16 cm long. The leaflets (3–10 cm) are elliptic-oblong, glabrous, finely toothed, and dark green above, pubescent beneath. The central leaflet is stalked, while the leaflets on either side are almost stalkless. During the early spring, delicate new shoots and unfurling leaves emerge a glorious chocolate bronze that is breathtaking against a blue sky. In Vancouver, fall colour produces a range of tints from yellow to pumpkin orange.

The flowers are borne in drooping axillary panicles up to 10 cm long during mid April through May. The individual flowers are up to 1.27 cm in diameter and have a delicate, milky fragrance. Clusters of 2 to 3 panicles are often grouped at branch nodes, creating a distinctive flowering pattern to the overall appearance of the plant. The flower colour in *Staphylea holocarpa* ranges from white to light rose. Our plant, of the variety *rosea* Rehder & E. H. Wilson, has flowers that are a spectacular coral-pink, as well as other characteristics of this variety, such as larger flowers, woolly indumentum on the leaf underside in early spring, and a subtle aquamarine leaf reminiscent of *Sorbus hupehensis*.

The fruits consist of membranous, pear-shaped, inflated capsules up to 5 cm long with globose grey seeds like buckshot. These capsules often remain on the plant throughout the winter. In the wild the species favours moist woodlands with full sun to partial shade.

HARDINESS: Zones 5–6. This subject is hardy in the Pacific Northwest and thrives in our benign climate. The early emergent growth of the species is unmolested by late frost, which is often the bane of this species in Western Europe. It appears to do well in the eastern United States as far north as Boston.

CULTIVATION: In cultivation it can be shaped into two different habits depending on preferred need. Maintained as a shrub, it makes a delightful specimen for a moderate-sized garden because of its bark, leaves, flowers, and general year-round appeal. It can also be trained as a small tree with formative pruning, and as such may hold promise for the future urban landscape. *Staphylea holocarpa* var. *rosea* is quite accommodating in terms of soil and moisture requirements and it remains free of pests and diseases. An evergreen tree or solid structure is ideal to silhouette the many ornamental charms of this species. A background setting in the company of other statuesque subjects, such as *Magnolia cylindrica* and *Photinia beauverdiana* Dashahe form, would work well.

PROPAGATION: The seeds are very hard and must not be allowed to dry out. They require an acid scarification followed by a 3-month cold stratification to break dormancy. Cuttings should be taken in early July.

Syringa sweginzowii Koehne & Lingelsheim 'Superba'
Chengdu lilac
Oleaceae   Plate 145

DISTRIBUTION: China, in Sichuan, at 2000 to 4000 m.

THE MARITIME climate of the Pacific Northwest Coast has not been traditionally regarded as optimal for lilacs. This sweeping statement after 25 years of observing and growing Asiatic species certainly requires clarification. The Chengdu lilac is one of the success stories in the David C. Lam Asian Garden, regularly producing abundant fragrant June blooms and displaying a complete lack of disease problems in our generally cool, moist springs.

It is likely that G. N. Potanin discovered this species in western Sichuan in 1893 as a member of a Russian Geographic Expedition (Fiala 1988, 72). Seed derived from this expedition established the plant in Europe and southern Russia by the turn of the century. It is conceivable that 'Superba' was raised by the Lemoines in France from seed that was collected by this expedition. It received an Award of Merit from the Royal Horticultural Society in 1918. The genus name *Syringa* was originally applied to *Philadelphus* (mock orange). It comes from the Greek *syrinx*, which means "a pipe," and refers to the hollow, pithy stems of that genus. The species was named by the German taxonomist Bernhard Adalbert Emil Koehne (1848–1918) in honor of Sweginow, governor of Livland (now Latvia and Estonia), in the early nineteenth century.

In cultivation in our garden, this elegant species makes a broad, open, fountain-shaped shrub, now 4.5 m high by as much across. The stems are reddish in youth, becoming purplish grey-brown with age. The outer branches become increasingly pendulous with age, especially in heavy flowering years. Our original plant came from the Hillier Nurseries of Winchester, England, in 1987. The papery leaves range from ovate-elliptic to lanceolate, 1.5–7 cm long by 1–4 cm wide, and are cuneate with an acuminate apex. The upper surface is shining, brownish green, and beneath, glabrous, with the leaf margins tinged red when first unfurling.

The flowers (7–25 cm long by 3–15 cm wide) are terminal or grow in lateral panicles that vary considerably in length and flower density. The calyx of individual flowers is campanulate with a narrow corolla tube up to 8 mm long, giving the individual flower clusters a lax appearance. Flower colour varies from pale pink to rosy lilac, with the interior of the corolla often being white. The cultivar

'Superba' is noted for the size of its pendulous flowers, though that may just represent part of a spectrum of natural variability. Our plant is vigorous and produces abundant, exquisite coral-pink paniculate flowers, displaying a dripping floral beauty of rare quality. The aromatic spicy fragrance of this species has an allure of its own, drawing the curious on a warm, calm evening.

In the wilds of Sichuan, E. H. Wilson recognized the quality of this plant. He collected seed in 1904 near Songpan (Sungpan) in the Min Shan region close to the border of Gansu Province. Hybridizers, including Victor Lemoine in France during the 1930s, and James Pringle of the Royal Botanic Gardens, Hamilton, Ontario, Canada, in the 1970s, created notable hybrids utilizing this species.

HARDINESS: Zones 5–7. This hardy species comes from the Tibetan borderlands of western Sichuan where intensely cold air pours off the Tibetan Plateau in winter. It is a lovely lilac that can be grown on both sides of North America with success, except in the southeastern United States where summer humidity and heat prevail.

CULTIVATION: This shrub excels in our garden, growing on the edge of a treed shrubbery underlain by a cool, moist, leafy soil, but it will tolerate bright sunny conditions for a good portion of the day during summer. Intelligent siting and integration with other shrubs of similar proclivities is key to displaying this glorious plant. Good associates include *Acer caudatifolium*, *Alnus firma*, and *Buddleja colvilei*, all having similar form and leaf morphology. The cool, moist springs of the Pacific Northwest can promote fungal and bacterial blights that afflict many lilacs, yet this species has proven to be extremely resistant to these problems in the Asian Garden. An open, sunny site and good air drainage are vital for this plant's success.

PROPAGATION: Seed dormancy is variable and often not long lasting. The species requires a 1- to 3-month cold period to encourage good germination. Cuttings taken in June give good results.

Syringa yunnanensis Franchet (SBEC 758)
Yunnan lilac
Oleaceae   Plates 146–147

DISTRIBUTION: China, in southwest Sichuan, southeast Xizang (Tibet), and northwest Yunnan, at 2000 to 3900 m.

SINCE THE 1980s, numerous collections of this species have been raised in the Pacific Northwest, with expectations that have been dashed by their growth habit and generally disappointing flowers. In general, it is an underwhelming species, with the notable exception of a remarkable form collected by the 1981 Sino-British Expedition to the Cang Shan (SBEC) from northcentral Yunnan and given to us by Roy Lancaster. Jean Marie Delavay documented *Syringa yunnanensis* in 1887, near Dali, but it was George Forrest who introduced this lilac to Western cultivation in 1907 from the same general location.

The plants we have raised from wild seed are tall, open branched, and willowy, with the upper branches becoming almost horizontal. These results are in sharp contrast to SBEC 758 which has a more robust appearance, with leaves and flowers to match, approaching the appearance of the related *Syringa villosa* from northeast China. Botanically it appears to be closer to the "ugly duckling" of the series *Villosae*, the Himalayan *S. emodi*. *Syringa yunnanensis* (SBEC 758) has grown into a large shrub of over 3 m with a stiff branching habit. The branches of young stems are light greyish brown with warty vertical lenticels. The mature bark exfoliates to form small, rusty brown, paper-like curls with vertical ridges along the stems.

The substantial leaves (16–22 cm long by 7–10 cm wide) are consistently ovate with acuminate apices and acute bases. The leaves are entire, dull green above, glaucous beneath, and margined with very fine hairs. Fall colour is not outstanding. The flowers are borne in erect, terminal panicles (5–18 cm long by 3–12 cm wide) that quickly assume a more horizontal or pendulous posture as the blossoms open and expand during late May. The flowers are lilac-red before opening and slowly fade to light purplish pink. The panicles, midway through their development, have an attractive bicoloured appearance and intense fragrance. On closer inspection of the flower corolla, prominent yellow anthers extend just to the corolla throat, a significant diagnostic feature for this species. In the wild, flower colour is quite variable, ranging from pale whitish purple to rose and lilac red.

George Forrest in the summer of 1906 reported seeing the purplish rose flowers of this species lighting up the margins of somber pine forests in the Cang Shan range. In 1981 the SBEC collected this species under SBEC 758 at the northern end of the Cang Shan, near Huadianba, at 2900 m. Roy Lancaster describes the site of this collection in his enthralling travelogue *Travels in China, a Plantsman's Paradise* (1989, 276). This plant is a first-rate introduction, distinctive in so many ways and in my opinion warranting at least varietal status. In the wild it often grows on limestone-derived soils, though it seems equally at home on the more acidic substrates of the Pacific Northwest.

HARDINESS: Zones (6)7b–8. This particular collection has survived a number of severe cold events since being planted in the David C. Lam Asian Garden in 1989, notably in December 1990 when the temperature reached −13.4°C. The relatively high elevation of the original collection probably explains this adaptability.

CULTIVATION: The shrub has grown with exceptional vigour in the Asian Garden, flowering heavily and being completely free of pests and diseases. Its dramatic presence, with glossy substantive foliage and striking flowers, helps it stand out from the rest of the lilac crowd. A large plant, it looks at home on the woodland edge in combination with striking shrubs such as *Hydrangea aspera* Villosa Group or the chestnut rose *Rosa roxburghii*. These species provide year-round appeal and complement each other admirably.

PROPAGATION: Cuttings should be taken in mid July and will root well within 2 months.

Xanthoceras sorbifolium Bunge
Yellowhorn, Official's hat tree
Sapindaceae   Plates 148–149

DISTRIBUTION: northern China.

THIS BUSHY TREE hails from the great arid arc of desolate country north of the Qin Ling Shan and eastward to the hills north of Beijing. It grows in a land of dense scrub that has been devoid of forests for millennia and experiences a severe continental climate. There, hardy native companions include *Caragana sinica*, *Cotinus coggygria* var. *cinerea*, and *Syringa pubescens*. Jean Pierre Armand David (1826–1900) introduced the shrub to the West by 1866, though it appears to have been introduced to the Crimea 40 years earlier by Russian traders and missionaries. The yellowhorn, or official's hat tree, is a survivor, growing moderately in youth and ultimately into a scrubby tree or large shrub to nearly 6 m tall. The main stems often lean, with an irregular, rugged, winter profile. The plant has the ability to sucker if the original plant fails. The distinctive compound leaves are 30 cm long, consist of 9 to 17 leaflets, and as the specific epithet implies, are similar to the leaves of *Sorbus aucuparia* (European mountain ash).

As the young leaves unfurl in late spring, the flowers are borne from axillary buds on the previous year's pithy stems. The showy racemes are 10–20 cm long, with individual flowers 2.5–3 cm across. The 5 white delicate petals offer striking contrast with the basal yellow or chartreuse stain that eventually fades to carmine, rose, or light mauve. Considerable variation in flower colour and size is an issue for plant breeders to consider. The generic name *Xanthoceras* comes from the Greek *xanthos*, meaning "yellow," and *keras*, meaning "horn," referring to the yellow horn-shaped growths between the petals.

The large pendulous fruits up to 6 cm long split to reveal several dark brown seeds reminiscent of *Aesculus*, the horse chestnut genus. The opened fruit has the appearance of a traditional Chinese official's hat. Seed production in the Pacific

Northwest is only likely to be reliable far inland, where seasonal heat and cold are ideal for this species. Nevertheless, the UBCBG plant produced 2 viable fruits for the first time in the summer of 2004. The flowers, leaves, and seeds are eaten cooked in northern China, but despite their sweet chestnut flavor, are still regarded by the Chinese as starvation food.

On the southern coast of British Columbia, this plant is suited to hot, dry, protected, rocky terrain where southern aspects and reflected heat combine to mimic its native land. It seems to thrive in dry, thin soils that are alkaline to moderately acidic.

HARDINESS: Zones (3)4–6(7). In the Pacific Northwest a microsite providing maximal heat, especially along the seacoast, will enhance plant maturation. The sequence of hot, dry summers has invigorated our plant. Interior regions of southern British Columbia to northern Oregon may provide ideal conditions for this continental species. On the eastern seaboard of the United States and west to states like Nebraska, it does well, but fails in the humid southern states.

CULTIVATION: At UBCBG, we received a plant from Forest Farm Nursery, Williams, Oregon, in 1991, after repeated failures from wild seed sources. This plant thrives on a sunny, south-facing slope. Here it flowers heavily in midsummer and is becoming an irregular, medium-sized shrub. It is an ideal plant for xeric patio gardens where concrete walls and structures ensure plentiful heat. It is a good companion for summer-flowering "dry-landers" such as *Carpenteria californica* (tree anemone), the delightful *Hippocrepis emerus* (scorpion senna), and *Koelreuteria paniculata* (goldenrain tree).

PROPAGATION: Seeds require no pretreatment, though a cold stratification for 2 to 3 months at 5°C hastens germination. Root cuttings taken in late winter with moderate bottom heat can be successful.

PART 3
Trees

Douglas Justice

Abies kawakamii (Hayata) T. Itō
Taiwan fir
Pinaceae
Plate 150

DISTRIBUTION: Taiwan.

FEW CONIFERS are as quick to become stately and refined as *Abies* species (true, or silver, firs). As a group, they tend toward symmetrical, conical growth and regular, whorled branching, and their aroma generally triggers deep-seated memories for many people. True firs are common Christmas trees in America: *A. balsamea* (balsam fir) in the Northeast; *A. fraseri* (Fraser fir) in the Southeast; *A. procera* (noble fir), *A. grandis* (grand fir), and *A. lasiocarpa* (subalpine fir) in the Far West. The evocative, subconscious appeal of such aromas is not limited to the firs, but it is a significant if underappreciated characteristic of these plants. Many firs are extraordinarily showy, with silver-backed needles and stout, violet-blue cones. Korean fir (*A. koreana*) is perhaps the most celebrated Asian species, as it both displays many of those qualities and is a size suitable for residential landscaping.

Alongside such relatives, *Abies kawakamii* stands out for its subtlety. Although gardeners tend to gravitate toward the nongreen firs—the startling blue-white needles of *A. concolor* (Colorado white fir), for example, make that species popular—they should not find it difficult to accept the green needles of Taiwan fir: they are richly glossy and dark green, flattened, slightly curved, and densely borne all around the shoot except for a broad, V-shaped channel on the upper side. Each branch gently sweeps away from the trunk and flattens out, exposing much of the needle surface to the light. These features are not only beautiful, but suggest shade tolerance and adaptation to a climate characterized by rain and wet snow. Indeed, *A.*

163

kawakamii is native to the high-altitude (3000 m), temperate rain forests of Taiwan on Mount Morrison (Yushan, the Jade Mountain), where it grows up to 35 m tall.

One of the most striking features of Taiwan fir is its bark. Even on relatively young plants, the bark is exceptionally attractive—pale, tawny brown, and flaking—which both contrasts with the deep green needles and reflects from within the layered branching. Eventually, the bark becomes corky. With age, *Abies kawakamii* produces attractive, purple, barrel-shaped cones. In the UBC garden, cool montane *Abies* species such as this one grow well and rapidly, holding their needles and lower branches for many years.

HARDINESS: This species is rated as hardy to Zone 6, but may not be successful outside maritime or montane regions where summer nighttime temperatures are excessive.

CULTIVATION: Like most other firs, *Abies kawakamii* requires rich, deep, well-drained soil as well as humidity and cool temperatures to thrive in a garden setting. Roots are sparsely produced and prone to breakage, so plants should be transplanted only when small and not disturbed unnecessarily.

PROPAGATION: Seed (when available) is easily germinated if sown when fresh or after a short period of cold stratification.

Acer morifolium Koidzumi
Mulberry-leaf maple
syn. *Acer capillipes* var. *morifolium* (Koidzumi) Hatusima
Sapindaceae   Plates 151–152

DISTRIBUTION: Southern Japan.

THE MULBERRY-LEAF maple is one of the least well known of all snake bark maples, yet, it may be one of the finest for garden culture, at least in those areas mild enough to support it. *Acer morifolium* is native to the islands of southern Japan where it grows into a small multistemmed spreading tree 10 to 15 m tall. Hardy at UBCBG, plants are exceptionally vigorous under moist conditions. The willowy shoots are initially glaucous purple-green, becoming faintly striped white on green, then fading to greenish brown where exposed to direct sunlight. Stems bear broad, nearly unlobed or 3-lobed, shiny leaves that hang vertically from the arching stems.

The leaves of *Acer morifolium* are reminiscent of a number of other snake bark maples—the much larger Japanese *A. capillipes* (the red-budded snake bark), for example, to which it is closely related. Both species display conspicuous, angular veins above, but on *A. morifolium*, blue-green cuticular wax effectively obscures the minor traces on the underside of the leaf. Waxiness hints at another close relationship, to the beautiful Formosan *A. rubescens*; but like *A. capillipes* that species is more upright and has leaves with more prominent lobing. Without doubt, though, the mulberry-leaf maple exceeds the others in the richness and subtlety of leaf colour. Emerging pinkish green, the leaves change in summer to grass green, or bronzy in full sun, and finally into shades of apricot, orange, and red before falling in autumn.

HARDINESS: Rated as hardy to Zone 8, but because this species leafs out in early spring, it is susceptible to damage from late spring frost.

CULTIVATION: The requirements of *Acer morifolium* are typical for snake bark maples: light shade, and acid, well-drained soil to which organic matter has been added. The thin bark must be protected from mechanical damage, such as by lawn mowers or line-trimmers, but also from full sun exposure. Plants are susceptible to winter sun damage (southwest injury) on cold, cloudless days. If the rays of the sun strike the frozen stems at right angles and cause living cells below the cork cambium to thaw, these can burst upon expansion. Such damage shows up as a vertical line of brown cork (bark), which may permanently disfigure the tree. Strategic shrub placement and retention of lower branches to create natural shading are effective preventative measures.

PROPAGATION: Like the seed of other snake bark maples, fresh *Acer morifolium* seed germinates well, and plants grow very quickly as seedlings. Maple species from the same section (snake barks are in the section *Macrantha*) tend to be promiscuous, so hybrids with other nearby snake barks will be common. Much garden-collected seed is hybrid, since only single individuals of species are generally planted and most require cross-pollination for reliable seed set. A good way to increase these plants is by stem cuttings, which are relatively easy to root, or by grafting onto seedlings of *A. davidii* or *A. capillipes*.

Acer schneiderianum Pax & K. Hoffmann
Schneider maple
syn: *Acer campbellii* subsp. *schneiderianum* (Pax & Hoffmann) E. Murray
Sapindaceae   Plate 153

DISTRIBUTION: Southwestern China.

A SIGNIFICANT number of maples basically unknown to Western gardeners and botanists have become available through a flurry of botanical exploration in China since the 1980s. The majority of these new arrivals are still under observation in Western botanical gardens and not yet commercially available; this is slowly changing, however, as access to and interest in these plants improves. Many of the finest species are *Acer palmatum* (common Japanese maple) relatives, and most share with that species its compact, spreading growth, fine branching, and diminutive leaves. Another feature that distinguishes the group (*Acer* section *Palmata*) is forked terminals: the paired buds on terminal branches nearly always result in a distinctive fork. Two species in section *Palmata* that have shown promise in the David C. Lam Asian Garden are *A. schneiderianum* and *A. wuyuanense*, both close relatives of more familiar species.

Acer schneiderianum has been considered by some to be a variety of or synonymous with *Acer campbellii*, a variable but mostly frost-tender species from the Himalayas and western China. In *A. schneiderianum*, however, plants exhibit fine branching and thin, broadly 3- to barely 5-lobed leaves; those of the stiffer, stouter *A. campbellii* are more leathery, with 5 to 7 wide, pleated lobes. The leaves of Schneider maple emerge glossy, bronze-green, becoming lighter green in summer, and changing to a clear, bright orange in autumn. Flowering is not particularly obvious—both the inflorescences and individual flowers are small—but the samaras can be showy. Plants from which our collections are derived displayed wide-spreading samara wings (to 5.5 cm) that were coloured pink to red. Whether cultivated plants retain this attractive trait remains to be seen, as our plants have not flowered.

Like the stems of many species in section *Palmata*, the stems of Schneider maple are photosynthetic and remain green for several years before they eventually become weathered and brown. They produce waxes that colour the younger stems bluish; wax-coated tissues are a common adaptation in plants growing at high elevation where there is significant exposure to ultraviolet light (the waxes are reflective and therefore protective). The species is native to montane areas of Yunnan and Sichuan, where it is reputed to reach only 6 to 8 m in height. In cultivation it forms an elegant, slender-stemmed tree.

HARDINESS: *Acer schneiderianum* is listed in Van Gelderen et al. (1994, 136) as hardy to Zone 6, but has not been trialled extensively in North America.

CULTIVATION: This species appears to be typical of section *Palmata* with respect to garden culture. It requires acid, well-drained, moisture-retentive soil and tolerates full sun exposure under moist soil conditions. Temperature fluctuations in the Pacific Northwest are probably similar to those found in the species's native habitat. Continental climate extremes and high nighttime temperatures may prove too great for this montane species.

PROPAGATION: *Acer palmatum* relatives are easily propagated from ripe seed that is harvested before the samara wings begin to turn brown. A period of cold stratification is necessary to overcome seed dormancy. Cuttings have also proven successful with a number of species and cultivars; however, nurseries typically graft species from this section onto *A. palmatum* seedlings.

Acer wuyuanense W. P. Fang & Y. T. Wu
Wuyuan maple, Chocolate maple
Sapindaceae   Plate 154

DISTRIBUTION: Southwestern China.

A SMALL PLANT, *Acer wuyuanense* is somewhat huskier than *A. schneiderianum*, with compact branching and more substantial leaves, but its habit is still spreading and graceful. Young stems are deep purple-green, often almost black, and in winter show up the lighter coloured buds and vertical flecks of developing bark. In cool spring weather, the sawtoothed, 5-lobed leaves emerge like oily, dark brown chicken feet, drooping from paired buds along the branches, the twisted lobes splaying slightly as they swell. Gradually becoming horizontal, they continue to expand and eventually lose their lustre before settling on a rich ivy green. At UBCBG, Wuyuan maple has grown to about 5 m tall by 3 m wide after 15 years. The species is reported to reach 7 m in its native Anhui and Jiangxi in mountains at 500 to 2000 m elevation. *Acer wuyuanense* has been linked with the more widespread *A. oliverianum* in the literature and may be a minor variant of that species, but the deep chocolate brown expanding leaves and inky stems of this collection from below Chi Guang Ge Temple (Merciful Light Pavilion) near Huang Shan, Anhui, suggest that it may be distinct.

HARDINESS: *Acer wuyuanense* is listed in Van Gelderen et al. (1994, 139) as hardy to Zone 6, but has not been trialled extensively in North America.

CULTIVATION: This species appears to be typical of section *Palmata* with respect to garden culture. Like *Acer schneiderianum*, it requires acid, well-drained, moisture-retentive soil and tolerates full sun exposure under moist soil conditions. Temperature fluctuations in the Pacific Northwest are probably similar to those found in the species's native habitat. Continental climate extremes and high nighttime temperatures may prove too great for this montane species, although *A. oliverianum* is known to be remarkably heat tolerant and, by extension, so might *A. wuyuanense*.

PROPAGATION: *Acer palmatum* relatives are easily propagated from ripe seed that is harvested before the samara wings begin to turn brown. A period of cold stratification is necessary to overcome seed dormancy. Cuttings have also proven successful with a number of species and cultivars; however, nurseries typically graft species from section *Palmata* onto *A. palmatum* seedlings.

Carpinus fangiana Hu
Monkeytail hornbeam
Betulaceae   Plates 155–157

DISTRIBUTION: Southwestern China.

HORNBEAMS (*Carpinus* species) are, on the whole, excellent urban trees, tolerating smog, drought, and poor soil conditions while doing an exceptional job of fending off pests and diseases. They are easily recognized by their smooth, usually unadorned bark, the small, alternately arranged, pinnately veined, serrated leaves that emerge from pointed buds, and small, nodding catkins composed of overlapping papery bracts. The name "hornbeam" derives from the Old English *horn*, meaning "hard," and *beam*, meaning "tree," alluding to the dense, heavy wood characteristic of the genus. Although hornbeams are generally marcescent (retaining leaves during dormancy), this juvenile characteristic is soon outgrown. Most gardeners are familiar with only a single hornbeam, the astoundingly common *C. betulus* 'Fastigiata' (fastigiate European hornbeam), but there are many more gardenworthy hornbeams in specialist and botanical garden collections that the gardening public has yet to know.

Of the three *Carpinus* species described in this book, *C. fangiana* is the least commonly available hornbeam in the West. This tree never fails to elicit a positive response, especially when its long, drooping catkins have formed—they are up to 30 cm long, straw white at first, then aging to tawny brown. The catkins are formed at the tips of the branches and are produced around the entire crown, reminiscent of Christmas tree garlands. The narrow bracts eventually reflex back as they dry out, exposing the attached nutlets. At this point in mid-autumn, handling or heavy weather tends to shatter the catkin and effectively disperse the seeds. The extraordinary leaves of *C. fangiana* are up to 20 cm long, emerging purple-bronze, changing to olive green, and marked with close-set, pinnate veins. They are the perfect foil for the tail-like catkins, as they, too, tend to hang vertically from the branchlets. Young shoots are dark brown or nearly black, with prominent raised white lenticels, while the trunk is dark brown and eventually somewhat scaly. In our garden, monkeytail hornbeam seems as adaptable as any, growing taller more quickly in shade, and denser where exposed to full sunlight.

HARDINESS: Zone 8. Asian hornbeams range in estimated hardiness from Zone 5 to Zone 8, although many of the published estimates derive from experience in Great Britain, where late spring frosts limit the survival of many plants. Cold-hardiness ratings for the genus are not well established in North America.

CULTIVATION: *Carpinus* is tolerant of most soils, except those that are either poorly drained or excessively open and gravelly. In the Pacific Northwest, irrigation is necessary in summer, since drought is a seasonal feature of the region.

PROPAGATION: Most *Carpinus* species are easily propagated from seed, although many have dormancy requirements that must be met, particularly if their seeds have dried out. Because individual trees may produce fruiting catkins regardless of whether flowers have been pollinated, the nutlets may not contain viable seed. Asexual propagation by stem cuttings has demonstrated variable success, but grafting onto *C. betulus* stock may be an option for harder-to-root species.

Carpinus tschonoskii Maximowicz
Chonosuke hornbeam
Betulaceae   Plate 158

DISTRIBUTION: Japan and Korea.

ANOTHER East Asian hornbeam species, but one more commonly associated with Japan where most seed of cultivated specimens is derived, is *Carpinus tschonoskii*. The habit of this species is decidedly wide spreading, with few, strong branches and with leaves more substantial than those of *C. turczaninowii* (to 9 cm); they are not unlike those of *Betula utilis* in size and shape. The parallel rows of deeply impressed veins cause the leaves to ripple and pucker slightly, which, together with the bristle-tipped, marginal teeth, adds reflectivity and a definite lightness to the trees. In late summer and fall, the fruiting catkins are eye-catching at first, with their showy reddish outer bracts, then upon ripening they appear like paper lanterns dotted about the crown. With age, the smooth steely bark is interrupted by lighter coloured, raised, longitudinal lines along the trunk and larger branches. Trees remain small and contained for many years, and overall, the impression is rather muscular, particularly in winter. The species name *tschonoskii* honours Chonosuke Sugawa, the celebrated Japanese collector who supported, and whose name was Russianized by, Russian botanist Carl Maximowicz. Chonosuke hornbeam ranges across Japan and Korea; Chinese plants under this name appear distinct and may be another species altogether.

HARDINESS: Zone 5 (estimated).

CULTIVATION: *Carpinus* is tolerant of most soils, except those that are either poorly drained or excessively open and gravelly. In the Pacific Northwest, irrigation is necessary in summer, since drought is a seasonal feature of the region

PROPAGATION: Most *Carpinus* species are easily propagated from seed, although many have dormancy requirements that must be met, particularly if their seeds have dried out. Because individual trees may produce fruiting catkins regardless of whether flowers have been pollinated, the nutlets may not contain viable seed. Asexual propagation by stem cuttings has demonstrated variable success, but grafting onto *C. betulus* stock may be an option for harder-to-root species.

Carpinus turczaninowii Hance
Turczaninow hornbeam
Betulaceae   Plate 159

DISTRIBUTION: Japan, northeastern China, and Korea.

ONE OF THE most distinctive hornbeams goes under the name of *Carpinus turczaninowii* (turch-an-nin-oh-vee-eye) in modern collections. Unfortunately, the plants we know may not actually be *C. turczaninowii*, as they do not fit the description of plants under that name collected earlier in the previous century. But Turczaninow hornbeam ranges from Japan and Korea across northern China, so it is possible that the species could show this degree of variation. The species name commemorates Nikolai Stepanovich Turczaninov (1796–1863), a Russian botanist whose writings and research shed much light on the flora of northern China. Original collections in the west from Gansu, China, are apparently upright, while recent collections from Sichuan differ in a number of characters, most notably in their graceful cascading habit. The small (5 cm), prominently veined leaves are maroon in spring, changing to deep green, and are borne on fine, wire-like branchlets. The bark is grey-brown and glossy. Russet and amber fall colours are typical for hornbeams.

HARDINESS: Zone 7 (estimated).

CULTIVATION: *Carpinus* is tolerant of most soils, except those that are either poorly drained or excessively open and gravelly. In the Pacific Northwest, irrigation is necessary in summer, since drought is a seasonal feature of the region

PROPAGATION: Most *Carpinus* species are easily propagated from seed, although many have dormancy requirements that must be met, particularly if their seeds have dried out. Because individual trees may produce fruiting catkins regardless of whether flowers have been pollinated, the nutlets may not contain viable seed. Asexual propagation by stem cuttings has demonstrated variable success, but grafting onto *C. betulus* stock may be an option for harder-to-root species.

Celtis choseniana Nakai
Chosen hackberry, Ullung nettle tree
Ulmaceae   Plates 160–161

DISTRIBUTION: Korea.

GENERALLY, the wet-winter, cool-summer conditions of the maritime Pacific Northwest are unsuitable for many plants that originate in warm summer climates. This pertains to the hackberries (*Celtis* spp.), a group of tough, mostly unremarkable elm relatives from the tropics and continental climate regions of the Northern Hemisphere. They are mostly large trees with deep roots and leaves adapted to heat and sunshine. Legion are the hackberry species that have been tried and have failed in the Pacific Northwest climate. Enter *C. choseniana*, a small Korean endemic species with a particularly handsome demeanour and the ability to tolerate a range of environments. This hackberry performs admirably in our garden, tolerating the cold shallow soil and November-to-March moisture as easily as it copes with the region's summer drought.

Celtis choseniana is a compact tree with a spreading scaffold of horizontally layered fan-like branches, ultimately forming a rounded or flattened irregular crown. The narrow, ovate shiny leaves are held horizontally (or slightly bent down) and alternate on short, zigzagging branches, providing a dramatic winter tracery. Autumn colour is poor—the leaves falling green or brownish—but the leaves do not linger on the branches. The hanging black fruits are reputed to be comparatively large (up to 12 mm in diameter), but plants at UBCBG have never flowered, despite being more than 15 years old. Material in cultivation was originally collected in the islands off northeastern South Korea in 1984. These plants were described as growing no more than 12 m tall, and their offspring will probably attain a similar stature. Chosen hackberry is one of few trees that deserves to be grown for its exceptional winter appearance alone.

HARDINESS: Probably hardy to Zone 7, but plants have not had extensive trialling in North America.

CULTIVATION: Hackberries are tough plants with a pair of Achilles heels: poor drainage and shade. Outside of these conditions, plants will perform admirably in almost any soil and exposure. Our plants in the open have a broadly rounded habit; in the shade they are spindly and unhealthy, despite deep, well-drained soil and summer irrigation.

PROPAGATION: Seed is easily sprouted after a long period of cold stratification. Both budding and stem cuttings have been successfully tried with other species and may be an alternative to seed, which may not be available.

Chionanthus retusus Lindley
Chinese fringe tree
Oleaceae   Plate 162

DISTRIBUTION: China, Japan, Korea, and Taiwan.

A SLOW-GROWING deciduous tree, *Chionanthus retusus* is similar to *C. virginicus* (American fringe tree) and equally deserving of wider planting. Fringe trees are so named because they produce foamy panicles of small flowers with strap-like white petals that droop attractively. In the American species, they are produced before leaf emergence and resemble the fringe on a military epaulette. Chinese fringe tree produces its flowers in smaller, more upright, rounded panicles—not as dramatically as in the heavier, droopier *C. virginicus*—but flowering is still showy at the ends of leafy branches in early summer. Beautiful indigo blue, olive-like drupes (single-seeded fleshy fruits) follow in autumn, and where summer heat is consistent, such as in continental climates (that is, away from coastal areas), fruit set can be spectacular. When ripe, the drupes are attractive to birds. Plants may be dioecious (with separate male and female individuals) or, at least, functionally dioecious. But single plants have been known to bear fruit, which suggests that the pollination biology of this species may be more complicated than it appears. Some people report that "male" plants have larger individual flowers and showier inflorescences, and this could be considered reasonable compensation for an absence of fruit, but more research is needed.

In youth, *Chionanthus retusus* is shrubby and its slow growth rate can be disheartening, but given optimal conditions plants eventually come into their own. The species makes flowers and fruit at a young age, however, and also produces interesting bark, a feature that is easily as striking and unusual as any of its floral attributes. Young shoots are covered with exfoliating, brown, papery bark and resemble the branches of *Acer griseum* (paper bark maple). As the stems age, however, dark zigzagging creases eventually coalesce and produce superb, grey-brown, heavily furrowed bark. The oppositely arranged, elliptic leaves are about 10 cm long, usually folded upward. Although late to emerge in spring, they remain on the branches well into November and are generally pest- and disease-free.

Chionanthus is derived from the Latin *chiono* (snowy white) and *anthos* (flower). The specific epithet *retusus* describes the retuse apex (not pointed, but rounded and notched) of some of the leaves. Despite being introduced to Britain by Robert Fortune in 1845, as well as consistently garnering high praise from all manner of past and present garden writers, it is still little known outside of botanical collections.

HARDINESS: Zone 6. As this species is widely distributed across Southeast Asia, provenance may determine cold hardiness to a considerable degree.

CULTIVATION: Too much protection from the elements renders plants weak and lanky. The best specimens are produced in deep, moist, well-drained soil, fully exposed to sun and wind. Trees are typically low branched or multistemmed and are known to grow to more than 20 m tall in the wild, but typical growth rates in the Pacific Northwest suggest a height of 4 or 5 m after about 20 years. Plants produce fibrous roots and are easily transplanted.

PROPAGATION: Chinese fringe trees in cultivation are produced from layers or, more commonly, from seed, which requires treatment to overcome double (seed coat and embryo) dormancy.

Cornus macrophylla Wallich
Large-leaf dogwood
Cornaceae   Plates 163–165

DISTRIBUTION: the Himalayas, China, and Japan.

IT IS DIFFICULT to successfully introduce trees that appear diminished in some respect to more spectacular relatives. Among tree aficionados, for example, *Cornus kousa* (Japanese dogwood) and *C. controversa* (giant or table dogwood) are currently acknowledged as the most desirable dogwoods for general planting in Western gardens. Both are relatively problem free and exceptionally attractive trees: *C. kousa* for its long-lasting blooms and striking, starburst habit, among myriad other superb qualities, and *C. controversa* primarily for its stunning, frosted wedding cake looks. While it is nearly impossible to identify serious shortcomings for either tree, it must be admitted that neither is small after 20 or 30 years or easily accommodated on a typical residential lot. Large-leaf dogwood, however, is well suited to such a situation and one never tires of its subtle good looks.

Large-leaf dogwood forms a small, often low-branched or multistemmed tree, usually no more than 8 or 10 m in height when grown in the open in gardens. The species is known to top 20 m in the wild, but such plants would have been heavily shaded and very old. In cultivation, branching is much like that of *Cornus controversa* (table dogwood), with strongly upright stems and horizontal tiers of branches. Each branch terminates with a handful of curving, shortly ascending twigs with prominent pointed buds. The grey-brown bark is smooth and attractively mottled in youth, eventually becoming plate-like with age.

True to its name, the leaves of *Cornus macrophylla* are large (about 17 cm long by 12 cm wide), and handsome, with prominent veins and slightly wavy edges. They are often creased down the midrib, making the lighter coloured underside visible. Creamy white flowers are individually tiny, but borne in broad, flattened corymbs at the tips of the branches. Blooming in June or July, they are highly visible against the fully expanded leaves. Once flowering has finished, small (6 mm), blue-black drupes form; cross-pollination is usually a prerequisite for fruit development.

HARDINESS: Zone 6.

CULTIVATION: This species is typical of the dogwood clan: it is easily grown in shade, or in sun with more or less constant moisture. Deep, well-drained soil is ideal, and growth rate and leaf size appear to be closely correlated with fertility. Some shading to the lower stems is advisable, as such thin-barked trees are subject to both winter injury and sunscald.

PROPAGATION: Seed, when available, is used as well as cuttings, which are exceptionally easy.

Cupressus torulosa D. Don
West Himalayan cypress
Cupressaceae   Plate 166

DISTRIBUTION: Himalayas.

ONE OF THE most striking of all small Asian conifers, west Himalayan cypress is a pyramidal or broadly columnar species with overlapping, horizontal, and eventually drooping branches, upswept at the tips. The slightly pendulous sprays of dark green foliage are reminiscent of some Lawson cypress (*Chamaecyparis lawsoniana*) cultivars, but the colour is more vivid and the foliage finer textured. Such a tree stands out in any garden situation, and Southern Hemisphere gardeners were quick to recognize this. The species and a number of novel cultivars are planted widely in Australia and New Zealand, where winter temperatures are not extreme. West Himalayan cypress is not generally considered reliably hardy, yet has done well in the Vancouver area for more than 10 years and probably deserves wider attention throughout the Pacific Northwest and the southern United States.

Cupressus torulosa is native to the western Himalayas from Nepal and northwest India to southwestern Xizang (Tibet) at between 1800 and 3300 m, where it is said to grow on limestone. It is apparently becoming endangered in habitat because of its high value as a timber tree. Trees in the wild grow to about 20 m, but in cultivation, probably less. While pyramidal in youth, older trees become softened and less angular, exposing more of the inner branches and trunk. Older stems have light brown bark that peels in vertical strips. The species is closely related to *C. cashmeriana*, the famous weeping Kashmir cypress, but is considerably hardier.

HARDINESS: Zone 7, but must have excellent drainage to survive wet winter conditions. The wide native range of the species suggests that hardiness is variable, depending on provenance.

CULTIVATION: At UBCBG, this tree grows quickly (6 m in 10 years) with summer irrigation. Survival in the cooler Pacific Northwest is not assured except in extremely well-drained soil. Like the majority of *Cupressus* species, this one requires aerated soil; anchorage may be compromised by wet conditions, either from root dieback or lack of root growth in areas of excess moisture. Amendments that raise soil pH (such as lime) may be beneficial, but trees appear to be healthy and vigorous in acid soils.

PROPAGATION: Seeds need a short period of cold stratification to overcome physiological dormancy, and then relatively dry conditions following germination. This species, like all other cypress, requires 2 years for the cones to ripen before the seeds are shed. Cuttings are the preferred propagation method for selected seedlings and cultivars, and are easy to root, as long as the cuttings are taken from juvenile growth.

Emmenopterys henryi Oliver
Rubiaceae   Plates 167–168

DISTRIBUTION: Southern and central China, northern Vietnam.

A RARE deciduous tree, *Emmenopterys henryi* is native to mountain valleys of southern and central China between 700 and 1300 m in elevation. Trees grow to 30 m tall in scattered pockets along river courses, although this species was probably more common in the mixed deciduous and broad-leaf evergreen mesophytic forest that once ran virtually unbroken across much of warm-temperate Southeast Asia. To a great extent, accessible territory in this region has now been deforested for timber, fuel, or agriculture. The tree is extraordinary for a number of reasons, not least that it has survived as long as it has. Many other species are similarly endangered, but *Emmenopterys* is among the most adaptable from this area.

In cultivation, *Emmenopterys henryi* is known primarily for its handsome, leathery, elliptic leaves (to 40 cm long), and its broad crown, which lend it an opulent, subtropical look, but it is also renowned for its frustrating reluctance to flower. Although introduced to the West nearly a century ago, few plants have bloomed. The largest is a multistemmed individual now 26 m tall (in 2003), planted in 1937 at Villa Taranto (Lake Maggiore, Italy), which started flowering in 1971. The first *Emmenopterys* reported to have flowered in North America is a 15-year-old tree at Silver Springs, Maryland.

At UBCBG and other gardens in the cooler Pacific Northwest, *Emmenopterys* grows quickly and shows the typical coppery new foliage, reddish petioles, and stiff horizontal branching that plants exhibit in warmer climates, though plants here have not yet become reproductive. The reason flowering is so eagerly anticipated is that the blooms are both unusual and beautiful. Inflorescences are about 15 cm tall, composed of yellowish white, trumpet-shaped flowers (2.5 cm long) that are borne with a few similarly coloured but eventually pink, persistent, showy bracts. The bracts, uncannily reminiscent of those on the climber *Schizophragma hydrangeoides*, are derived from the calyx (sepal) tissues of outer flowers (as they are in the unrelated *Schizophragma*). The genus name *Emmenopterys* is from the Greek, *emmeno*, meaning "persist," and *pteryx*, meaning "bract," while the epithet *henryi* commemorates the great Irish plant explorer Augustine Henry (1857–1930).

HARDINESS: Zone 7 or 6. Plants appear to gain cold hardiness as they mature. Young plants may be frozen out with only a few degrees of frost.

CULTIVATION: Established trees require protection from cold winds, and from frost in youth. *Emmenopterys* responds well to woodland conditions (partial shade and moist soil, rich in organic matter) with good drainage and is said to tolerate calcium-rich, high-pH soils. More sunshine is probably required to ripen wood adequately and initiate flowering, but leaves are larger and more majestic in shade. However, as this is an emergent species, in heavy shade stems tend to shoot up unbranched until they reach an opening in the canopy (and more light), before producing significant scaffold branches and flowering.

PROPAGATION: Plants are relatively easy to propagate from seed, which is seldom available. Cutting propagation is evidently readily accomplished using material from root suckers.

Fraxinus sikkimensis (Lingelsheim) Handel-Mazzetti
Sikkim ash
syns. *Fraxinus paxiana* var. *sikkimensis* Lingelsheim, *F. suaveolens* W. W. Smith
Oleaceae   Plates 169–170

DISTRIBUTION: Northeastern India and western China

THIS HANDSOME ash is native to forested areas at 2000 to 3000 m elevation from the eastern Himalayan Indian states of Assam and Sikkim to the western Chinese provinces of Sichuan, Xizang (Tibet), and Yunnan. Trees are strong growing, with silvery green 4-angled shoots, and light purplish brown buds; the terminal buds are often large and attractive. In nature, Sikkim ash is known to grow up to 17 m tall and form a rounded crown. The species belongs to the flowering ashes, a group known primarily by *Fraxinus ornus* (manna ash), whose fragrant flowers, although individually small, have strap-like, creamy white petals that are borne profusely in large, blousey inflorescences after the leaves have emerged. Those of *F. sikkimensis* are considered among the showiest of the genus, being yellow-white and produced in May or June in large panicles 15–30 cm long at the tips and in the axils of the leafy shoots around the entire crown.

Fraxinus sikkimensis is a close relative of *F. paxiana* and differs primarily in its higher elevation habitat and more western range, its smaller overall size, strictly

Plate 150. A 20-year-old *Abies kawakamii* (Taiwan fir). Photo by June West

Plate 151. The juvenile leaves of *Acer morifolium* (mulberry-leaf maple), similar to those of *A. capillipes*. Photo by June West

Plate 152. A hint of autumn colour in *Acer morifolium*. Photo by Douglas Justice

Plate 153. The smooth, glossy, young leaves of *Acer schneiderianum* set it apart from other Japanese maple relatives. Photo by Douglas Justice

Plate 154. *Acer wuyuanense* (Wuyuan maple). Photo by Douglas Justice

Plate 155. A young *Carpinus fangiana* (monkeytail hornbeam), showing the dense, rounded form of this tree in the open. Photo by Peter Wharton

Plate 157. *Carpinus tschonoskii.* Photo by Gerald Straley

Plate 156. Emerging leaves of *Carpinus fangiana.* Photo by Gerald Straley

Plate 159. The pendulous fruit clusters (monkey tails) of *Carpinus fangiana.* Photo by Gerald Straley

Plate 158. *Carpinus tschonoskii* (Chonosuke hornbeam) exhibiting spreading, silvery, muscular stems. Photo by Douglas Justice

Plate 160. Lichen on *Celtis choseniana* (Chosen hackberry), encouraged by the clean, maritime air at UBCBG. Photo by Douglas Justice

Plate 161. *Celtis choseniana.* Photo by June West

Plate 162. The showy, early summer flowers and graceful habit of *Chionanthus retusus* (Chinese fringe tree). Photo by June West

Plate 163. Fruiting branches of *Cornus macrophylla* (large-leaf dogwood). Photo by Douglas Justice

Left: Plate 164. The smooth bark of young *Cornus macrophylla*, an excellent substrate for the growth of lichens. Photo by Douglas Justice

Above: Plate 165. The glossy leaves and creamy white flowers of *Cornus macrophylla*. Photo courtesy of Global Book Publishing Photo Library

Plate 167. The lush growth typical of *Emmenopterys henryi* in shaded woodland conditions. Photo by June West

Plate 166. The striking, upright, conical form of *Cupressus torulosa*. Photo courtesy of Global Book Publishing Photo Library

Plate 168. *Emmenopterys henryi*, along the Shi Ba Han Lo (18-Curve Pathway), Huangshan Mountain, Anhui Province. Photo by Peter Wharton

Plate 169. *Fraxinus sikkimensis* (Sikkim ash). Photo by June West

Plate 170. A senescent leaf of *Fraxinus sikkimensis* on the first day of November. Photo by Daniel Mosquin

Left: Plate 171. *Juglans cathayensis* (Chinese walnut) at 20 years, in the VanDusen Botanical Garden, Vancouver, British Columbia. Photo by June West

Above: Plate 172. Emerging leaves and developing fruit on *Juglans cathayensis.* Photo by June West

Plate 173. Chinese walnuts, the sweet, edible fruit of *Juglans cathayensis.* Photo courtesy of Global Book Publishing Photo Library

Plate 174. Young stems of *Lagerstroemia fauriei* (Yakushima crape myrtle). Photo by June West

Plate 175. *Larix griffithii* (Sikkim larch). Photo by June West

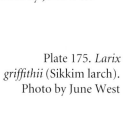

Left: Plate 176. The yellow cone form of *Larix griffithii* in May. Photo by June West

Right: Plate 177. The red cone form of *Larix griffithii* in May. Photo by June West

Left: Plate 178. *Lindera erythrocarpa* (red berry spicebush), in flower in early spring. Photo by June West

Above: Plate 179. The delicate leaves of *Liquidambar acalycina* (Chinese sweetgum). Photo by June West

Plate 180. A young *Lithocarpus variolosus* (varied leaf tanoak). Photo by June West

Plate 181. New growth flush on *Lithocarpus variolosus.* Photo by Peter Wharton

Plate 182. *Malus yunnanensis* (Yunnan crabapple). Photo by June West

Plate 183. The leaves of *Meliosma pinnata* var. *oldhamii* (Oldham worm-head tree). Photo by June West

Plate 184. The exuberant growth of *Meliosma veitchiorum* (Veitch worm-head tree). Photo by Peter Wharton

Left: Plate 185. *Meliosma veitchiorum.* Photo by June West

Above: Plate 186. A 10-year-old *Melliodendron xylocarpum* (Chinese parasol storax). Photo by June West

Plate 187. *Melliodendron xylocarpum* flowering in March. Photo by June West

Above right:
Plate 188. New growth of *Nyssa sinensis* (Chinese tupelo). Photo by Douglas Justice

Right:
Plate 189. Autumn colour on *Nyssa sinensis*, at VanDusen Botanical Garden, Vancouver, British Columbia. Photo by June West

Left: Plate 190. *Photinia beauverdiana* in full flower, late spring. Photo courtesy of Global Book Publishing Photo Library

Above: Plate 191. Plants derived from the Dashahe Cathaya Reserve in China, displaying larger fruits than other selections of *Photinia beauverdiana.* Photo by June West

Left: Plate 192. A young *Picea smithiana* (Morinda spruce) in VanDusen Botanical Garden, Vancouver, British Columbia. Photo by June West

Right: Plate 193. *Picrasma quassioides* (picrasma, or bitter ash). Photo by June West

Plate 194. The colourful leaves and fruit of *Picrasma quassioides*. Photo by Mark Flanagan

Plate 195. *Platycarya strobilacea* (cone walnut). Photo by June West

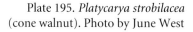

Plate 196. Male catkins of *Platycarya strobilacea*. Photo by June West

Plate 197. Infructescences of *Platycarya strobilacea* in early autumn. Photo by Tony Kirkham

Plate 198. *Quercus acuta* (Japanese evergreen oak), Photo by June West

Plate 199. *Quercus myrsinifolia* (bamboo leaf oak) with *Rhododendron calophytum* and *R. lutsescens.* Photo by June West

Plate 200. The characteristically smooth bark of *Quercus myrsinifolia.* Photo by June West

Plate 201. The large flowers of *Rehderodendron macrocarpum* (Rehder sausage tree) are borne, styrax-like, along the branches' undersides. Photo by June West

Plate 202. Fruits of *Rehderodendron macrocarpum.* Photo by June West

Left: Plate 203. A young *Sorbus commixta* (Japanese mountain ash). Photo by June West

Above: Plate 204. The leaves and flowers of *Sorbus commixta.* Photo by June West

Plate 205. The extraordinarily glossy, vermilion fruits of the Mount Kurohime form of *Sorbus commixta*. Photo by June West

Plate 206. *Sorbus hupehensis* (Hubei mountain ash). Photo by June West

Plate 207. Autumn colour of *Sorbus hupehensis* at VanDusen Botanical Garden, Vancouver, British Columbia. Photo by Gerald Straley

Left: Plate 208. *Sorbus oligodonta* (kite-leaf rowan) displaying the rounded crown typical of the species. Photo by June West

Above: Plate 209. *Sorbus oligodonta* berries in September. Photo by Judy Newton

Left: Plate 210. *Sorbus yuana* (Yuan whitebeam). Photo by Judy Newton

Above: Plate 211. The flowers of *Sorbus yuana.* Photo by Peter Wharton

Plate 212. The foliage and berries of *Sorbus yuana* in late autumn. Photo by Judy Newton

Plate 213. The dainty bells of *Styrax hemsleyanum* (Hemsley snowbell). Photo by Douglas Justice

Left: Plate 214. *Taiwania cryptomerioides* (coffin tree) at 10 years. Photo by Gerald Straley

Above: Plate 215. The juvenile foliage of *Taiwania cryptomerioides*. Photo by Gerald Straley

Plate 216. *Tetradium glabrifolium* (Hong Kong evodia). Photo by June West

Plate 217. *Tetradium glabrifolium*, with flowers in October. Photo by June West

Plate 218. *Tetradium hupehensis* (Hubei evodia, or Chinese bee tree). Photo by Mark Flanagan

unisexual flowers, and felted leaf underside, but these differences are considered by many to be too slight to differentiate them as separate species. In any case, the plants cultivated in the David C. Lam Asian Garden are lush growers, with beautiful, prominently veined, elder-like leaves; they are much different from the less robust, leaner specimens of *F. paxiana* nearby. Our plants have not yet flowered.

HARDINESS: Zone 5. With such a wide-ranging species, provenance may determine cold-hardiness to some extent. Because the UBCBG plants are from seed collected in northern Yunnan at over 2900 m, they should be very hardy.

CULTIVATION: Trees in the genus *Fraxinus* are extremely adaptable, noted for their tolerance of atmospheric pollution, wind, and poor soils. In our garden, Sikkim ash appears unperturbed by poor drainage, shade, and drought.

PROPAGATION: *Fraxinus* seed usually requires a period of cold stratification to overcome dormancy. Sikkim ash could also presumably be budded onto seedlings of *F. ornus*.

Juglans cathayensis Dode
Chinese walnut
Juglandaceae   Plates 171–173

DISTRIBUTION: China.

WALNUT TREES are known for their open, broadly spreading crowns and large leaves, and Chinese walnut is an exceptional example of this form. *Juglans cathayensis* is celebrated not only for its imposing structure, including its huge spreading branches, barely punctuated with lateral shoots, but for its tent-like leaves and silvery bark, reminiscent of the more familiar North American *J. cinerea* (butternut). Stouter and less congested than either butternut or the closely related *J. ailanthifolia* (Japanese walnut), Chinese walnut is also better adapted to shade and moist soil conditions. Native to a broad swath of central and western China and Taiwan (as an element of the mixed mesophytic forest, it grows up to 25 m tall in moist areas. In cultivation, its long branches are matched by spreading roots, making this a tree suitable only for large spaces, especially close to water.

Early on, the leaves, young branches, and developing fruits are covered in viscid hairs, velvety maroon in colour. The stickiness soon subsides and ultimately only the leaves retain a discernable fuzz resulting from short, stellate hairs that are silvery white and particularly close-set on the leaflet underside. Catkins are long (up to 40 cm), as are the pendent stems that hold clusters of 5 to 10 oily nuts. Reports of the nuts (our plants have not fruited) suggest that they are tasty but difficult to extract from the thick rind and extra hard shell.

HARDINESS: Although not as winter-hardy as the closely related Manchurian walnut (*Juglans mandschurica*), Chinese walnut is known to survive in Zone 6 or Zone 5. Plants are comparatively late to leaf out and generally unaffected by late spring frost.

CULTIVATION: Given an ideal site, with deep, well-drained, rich soil and some protection from wind, this species grows with exceptional vigour. At UBC a plant derived from seed from Beijing Botanical Garden is now 8 m tall by 12 m across after only 20 years.

PROPAGATION: Seeds (walnuts), extracted from their husks, require a 3–4 month period of cold stratification to overcome embryo dormancy before they will germinate.

Lagerstroemia fauriei Koehne
Yakushima crape myrtle
syn. *Lagerstroemia subcostata* var. *fauriei* (Koehne) Hatusima ex T. Yahara
Lythraceae   Plate 174

DISTRIBUTION: Southern Japan.

LAGERSTROEMIAS are noted for spectacular flowers and interesting peeling and flaking bark. In warmer parts of North America, they are hardly unusual, but the crape myrtles of commerce (primarily *Lagerstroemia indica* selections and their hybrids with *L. fauriei*) are difficult to grow in cooler parts of the Pacific Northwest. Vancouver's climate is characterized by cool, wet winters followed by long, mild spells in spring and not enough summer heat for adequate ripening of many warm climate–derived plants. Fortunately, Yakushima crape myrtle is both beautiful and surprisingly tough in this region. Flowers are white and individually small, but are fragrant and produced in showy, terminal panicles in summer. Admittedly more modest in flower than other lagerstroemias, this species easily compensates for any floral deficiency with its elegant, restrained habit, flaking bark, and ease of cultivation.

Lagerstroemia fauriei is native to Yakushima Island in southern Japan where it is a rare tree. The privet-like, forward-pointing lanceolate leaves are glossy dark green and held in 2 neat, subopposite ranks along the thin shoots. They remain virtually unharmed by pests or disease throughout the season. In autumn, the leaves turn shades of yellow and russet before they fall. Then, we anticipate with excitement the winter bark effect, which is one of the tree's finest features. Over much of the plant, the bark flakes and peels in strips and patches, exposing warm red, brown, and tan underlayers. The effect is superb. At UBCBG, Yakushima crape myrtle is slow growing, eventually forming a slender, multistemmed, open-branched tree of 5 to 8 m tall. In warmer climates, growth is considerably more rapid and lush, resulting in a spreading, bushy habit to 10 m or so in height.

Only a few cultivars of *Lagerstroemia fauriei* are in commerce. 'Fantasy' and 'Townhouse', both with reddish stems and small flowers, were released by the J. C. Raulston Arboretum of North Carolina State University, while the larger-flowered 'Kiowa' was introduced by the U.S. National Arboretum. A fourth cultivar ascribed to *L. fauriei*, 'Sarah's Favourite', has very large flowers and stronger fall colouration, and is probably a hybrid. There appears to be considerable variation within Yakushima crape myrtle and much potential for selection. The single plant at UBC was derived from original germplasm of the J.C. Raulston Arboretum, although its bark characteristics differ markedly from the cultivars selected there.

HARDINESS: Zone 6, but excess winter moisture would eliminate any significant frost tolerance.

CULTIVATION: Setting is extremely important for *Lagerstroemia*. In areas where flowers are possible, full sun and moisture-retentive soils are needed for best flowering. Where flowering is marginal, those site factors are essential, but reflected heat, such as from pavement or masonry walls, is a significant benefit. In warm-climate areas, late-summer extension growth is sometimes not hardened sufficiently to survive the winter. In the Pacific Northwest, summer drought generally restricts late-summer growth.

PROPAGATION: Lagerstroemias are noted both for their ease of propagation from seed, which germinates quickly without pretreatment, and for their rapid growth as seedlings. They are also easily rooted from softwood and hardwood cuttings.

Larix griffithii Hooker f.
Sikkim larch
syn. *Larix griffithiana* (Lindley & Gordon) Carrière
Pinaceae   Plates 175–177

DISTRIBUTION: Eastern Himalayas, from eastern Nepal to Bhutan and Sikkim and Arunachal Pradesh in India.

THE GENERAL public variously tolerates deciduous conifers. In cold climates they are grown because of lack of available variety in plants generally, and in temperate gardens often for their venerable lineage, such as the primitive and exotic pair *Metasequoia glyptostroboides* (dawn redwood) and *Taxodium distichum* (bald cypress). The benefit of a deciduous conifer in the landscape should not be overlooked, particularly in a dull climate. As for any deciduous tree, it provides more light on the garden floor in winter. By far, the most common deciduous conifers in the Northern Hemisphere are the larches, but since they are, with few exceptions, a scruffy lot, they are generally a much-maligned group. *Larix griffithii* is an exception.

Individual needles of Sikkim larch are long, and among individuals they range from green to steel grey. The tree's habit is regular, narrow, and tall, with pendulous, golden brown branches somewhat reminiscent of *Larix occidentalis* (western larch), from the mountains of southern British Columbia and the northwestern United States. Like that species, the autumn colour of *L. griffithii* is pure, flaxen yellow. Sikkim larch cones are perhaps the largest of all *Larix* species (9 to 15 cm long) and among its most interesting and ornamental features. They are purple-brown with reflexed scales and are held stiffly upright when ripe. The young, developing cones are extraordinarily large and are coloured either chartreuse or red-magenta. Both forms are exceedingly showy in the spring.

Sikkim larch is native to the high subalpine regions of the eastern Himalayas. Specimens at UBCBG were grown from seed collected in Bhutan by Keith Rushforth. The closely related *Larix griffithii* var. *speciosa* (syn. *L. speciosa*) is found in northwestern Yunnan and southeastern Xizang Zizhang (Tibetan Himalayas). That variety differs primarily in the broader needles and shorter cones.

HARDINESS: *Larix griffithii* is listed in British publications as hardy to Zone 7, but this is probably overly conservative and reflects the problems gardeners in that region have with late-spring frosts on plants with early emerging growth. In areas with clearly defined seasons, the species should prove much hardier.

CULTIVATION: Sikkim larch appears to grow well in a variety of garden situations, from dry, infertile, hard-packed soil to richer, even soggy conditions. Like most conifers, it benefts from good air circulation and full exposure to the sun, especially the bluer needled forms.

PROPAGATION: Seed germinates easily and is abundantly produced in the large cones, but seed viability may be low from isolated individuals.

Lindera erythrocarpa Makino
Red berry spicebush
syn. *Lindera thunbergii* Makino
Lauraceae   Plate 178

DISTRIBUTION: Eastern China, Korea, and Japan.

LINDERAS are generally not well known outside of botanical collections or their native haunts, which is unfortunate, because several are excellent garden plants.

There are three eastern American species, including *Lindera benzoin* (spicebush), which is grown as an ornamental, and is sometimes used for the stimulant, tonic, and diaphoretic properties of its bark. The others are tropical and temperate species native to Asia (there is a single species native to Australia). All have aromatic bark, flowers, and foliage, and some are evergreen. While many are ornamental, most are too tender for temperate gardens. Some exceptions include the shrubby *L. obtusiloba* and even the less well-known *L. erythrocarpa*.

Lindera erythrocarpa is a vigorous, deciduous woodlander from eastern China, Japan, and Korea noted for its pest-free, willow-like leaves, smooth clean stems, upright stature (6 to 15 m tall), and tiered branching. The spicy flowers are bright yellow and very showy despite their small size. They are produced early in the spring before the foliage emerges, with an effect reminiscent of the flowering of *Cornus mas* (Cornelian cherry); but red berry spicebush is more elegant and tree-like in habit. The alternately arranged, oblanceolate leaves emerge in early May, some weeks after flowering is complete. Copiously produced along the sweeping branches, they are glossy, deep green above, and glaucous below. The specific epithet *erythrocarpa* refers to the berries (technically, drupes), which are red. To be accurate, red is a transitional fruit colour following yellow, and preceding purple, and ultimately black. All berry colours may be present on the plant during the late summer and fall, but like deciduous hollies, male and female plants must be present for berry set. Fruit production probably also requires significant summer heat, which is lacking at UBC and in other maritime climates. Even without fruit, these plants are outstanding ornamentals.

HARDINESS: Zone 6.

CULTIVATION: Linderas prefer lime-free soil that is well drained and moisture retentive, though established plants tolerate drought. The laurel family is noted for its arsenal of endogenous antipest phytochemicals, indicated by the lemon and ginger aroma of their leaves and inner bark. Not surprisingly, pest and disease problems with linderas are rare. Plants in regions where spicebush is native, though, may suffer from adapted resident pests. Plants grow quickly in moist soil with light shade in the cooler Pacific Northwest, but also grow well in the open, both here and in areas with high summer temperatures. Plants are shrubbier in the open, but will still eventually form small trees; however, they may be as wide as they are tall.

PROPAGATION: Linderas are usually grown from seed that requires a period of cold stratification to overcome dormancy, but they are also propagated easily from cuttings and layers.

Liquidambar acalycina H. T. Chang
Chinese sweetgum
Altingiaceae   Plate 179

DISTRIBUTION: Southern China.

CHINESE SWEETGUM is a striking, irregularly branched tree. Its leaves are broad, each of its 3 lobes more generous and irregular than any other sweetgum species. New leaves emerge a remarkable deep burgundy-purple, which makes the trees at UBCBG a significant visitor attraction. Eventually fading to bronze then dark green, the leaves have an attractive velvety lustre; they sit languidly in clusters astride the long, horizontal branches. *Liquidambar acalycina* is somewhat similar in form to the tenderer *L. formosana*, and the two species have overlapping ranges in southern China, but *L. acalycina* is native to higher elevations in montane evergreen forests, where it grows to 25 m in height. The species looks to be somewhat smaller in cultivation. Still, it revels in our garden and colours superbly orange and deep purple in autumn. Chinese sweetgum sheds its leaves reliably before the winter solstice, whereas plants of the more commonly cultivated *L. formosana* usually hold their leaves well into the new year at UBCBG. Our largest tree is derived from seed collected on the 1980 Sino-American Botanical Expedition to Hubei Province from the *Metasequoia* region of Lichuan Xian at 1500 m elevation.

Liquidambar is familiar to many as the common name for the North American *Liquidambar styraciflua*, a species somewhat reviled for its habit of dropping copious, walnut-sized, spiny "gumballs" along city streets where it is commonly planted. The name itself comes from Latin *liquidus*, meaning "liquid," and *ambar*, meaning "amber," referring to the fragrant resin, liquid storax, derived from the sap of some species. In our garden, plants of *L. acalycina* do not appear to flower, possibly because of the absence of significant summer heat to adequately ripen wood. If they were to flower and fruit, however, they would be less of an issue for pedestrians than their American counterpart, as the gumballs produced are smaller and spongier. The epithet *acalycina* (Latin *a* for "without" and *calycina* for "calyx") hints at the reduced calyx of the female flowers compared with the related *L. formosana*; however, the floral anatomy of liquidambars in general is not obvious to the casual observer.

HARDINESS: Zone 6; but seed from more southerly collections in China may not prove as cold-hardy.

CULTIVATION: This species is of easy culture, preferring moisture and rich soil, but also tolerating drier conditions. In moderate climates in moist soil, growth continues late in the season and may not harden off adequately, which contributes to overall weakness and increases susceptibility to breakage during winter storms.

PROPAGATION: Seed, if available, should be sown when ripe. Individual trees may not set viable seed.

Lithocarpus cleistocarpus (Seemen) Rehder & E. H. Wilson
Closed cup tanoak
Fagaceae

DISTRIBUTION: Southeast and central China.

MUCH LIKE evergreen oaks in durability and stature, the Asian tanoaks (also known as lithocarps) differ in their upright catkins, thick, bony fruit wall, and untoothed leaves. Tanoaks are excellent trees for hot, dry climates, but also appear equally happy in the winter-wet, summer-dry climate of the Pacific Northwest, where they are slower in growth and make smaller trees. The lovely Japanese tanoak (*Lithocarpus edulis*) is currently in commercial production on the West Coast, but few others are cultivated, even in botanical collections. Happily, recent expeditions to China have brought back a number of species, and among these, *L. cleistocarpus* and *L. variolosus* appear to be exceptionally garden worthy. Leaf emergence with tanoaks is an extraordinary sight, as the leaves are generally covered in silky reflective hairs and unfold in large, upright clusters, the leaf blades drooping softly, then gradually expanding as the hairs on the upper surface disappear.

Closed cup tanoak is a columnar evergreen with dense, broadly ascending branches. Its dark green, silver-backed leaves are luxuriant and enormous (often to 30 cm long). They are particularly striking as they emerge in early spring, like handfuls of creamy silver flags. Fully developed, the wide, wavy-margined leaves seem to droop like elephant ears along the branches. They persist for 4 or 5 years and seem impossibly outsized on young plants. Older trees are rare in the West, and our experience with this species is limited to its performance in the David C. Lam Asian Garden, which can be described in two words—trouble free and robust. Chinese tanoak will probably retain its narrow crown, and although published accounts list the tree as growing only to 10 to 15 m, it will undoubtedly exceed that in the Pacific Northwest, as it has nearly reached 8 m in 12 years. The species name *cleistocarpus* comes from the Latin meaning "closed fruit" and refers to the cupule of this species, which almost completely encloses the nut.

HARDINESS: Zone 7.

CULTIVATION: *Lithocarpus cleistocarpus* appears to be completely happy in the UBC garden in dry, poor soil and in sun or shade, but plants are more compact in the open. It is early into growth in spring, which may preclude growing it successfully in areas that experience late frosts.

PROPAGATION: Easily germinated from seed, many *Lithocarpus* species are also successfully propagated by semiripe-wood cuttings.

Lithocarpus variolosus (Franchet) Chun
Varied leaf tanoak
Fagaceae   Plates 180–181

DISTRIBUTION: Southwest China and Vietnam.

VARIED LEAF tanoak is one of the most exciting tree introductions of recent years. It has a limited range in the mountains of subtropical southwestern Sichuan, northwestern Yunnan, and adjacent Vietnam, but it is a subalpine species and appears to be relatively cold-hardy (our plants were collected at 3100 m). Plants have an open pyramidal habit in youth, the branches being borne horizontally from a central bole, much like a young *Quercus palustris* (pin oak) in miniature. The glossy evergreen leaves are wavy edged and cupped slightly so that their silver-grey backs are displayed to advantage, particularly under breezy conditions. Typically around 10 cm long and narrowly ovate but variable in size and shape, the leaves are exceptionally blemish free and long lived and are one of this species's most outstanding characteristics. Flushing in spring and again in summer, the new leaves are tawny silver and showy. *Lithocarpus variolosus* produces catkins and fruit at an early age, and the fruit clusters, though somewhat oddly shaped (like partially molten extruded plastic), are not an obtrusive feature. Like the closed cup tanoak, this species is not well characterized in cultivation, but given its juvenile form and moderate rate of growth in our garden, it looks to form a broad-headed tree of perhaps 10 m height.

HARDINESS: Zone 7. *Lithocarpus variolosus* is probably hardier than *L. cleistocarpus*.

CULTIVATION: Like *Lithocarpus cleistocarpus*, *L. variolosus* appears to be completely happy at UBCBG in dry, poor soil and in sun or shade, but plants are more compact in the open. Varied leaf tanoak occurs on limestone in nature, but is perfectly adapted to the acid soils at UBCBG. It is early into growth in spring, which may preclude growing it successfully in areas that experience late frosts.

PROPAGATION: Easily germinated from seed, many *Lithocarpus* species are also successfully propagated by semiripe-wood cuttings.

Malus yunnanensis (Franchet) C. K. Schneider
Yunnan crabapple
Rosaceae   Plate 182

DISTRIBUTION: China

In the cool, rainy climate of the Pacific Northwest, cultivated crabapples are frequently diseased, unattractive, and short-lived. Yunnan crabapple is a montane, forest-edge species native to western China and one of the few *Malus* species that grows well under the coastal regime. It deserves to be grown for other reasons, as well. Ultimately of small size (less than 10 m) with a broad, low crown, its bears large, felted, heart-shaped leaves that are handsome, turning yellow, orange, or scarlet in autumn. In early spring, white, 1.5-cm flowers from pinkish buds are produced in large corymbs. Following pollination, the resultant 3-cm-diameter crabapples can be yellow, red with white spots, or entirely russet. The pomes hang in small bunches on long peduncles below the horizontal branches.

The most obvious difference between this tree and most other *Malus* species is that Yunnan crabapple remains exceptionally unblemished throughout the growing season. Scab can devastate both the foliage and fruit of *Malus* species at UBCBG, but this one appears consistently resistant throughout, which is a welcome feature in a wet climate. *Malus yunnanensis* also tolerates a variety of garden situations. Our accessions are planted in both poor, dry soil and in wet ground. One extraordinary individual is growing strongly and producing fruit despite being shaded for more than half the day. The cultivar 'Veitchii' is sometimes available in the United Kingdom, differing in its more upright branching and darker fruit colour.

HARDINESS: Zone 6 or colder.

CULTIVATION: *Malus* species prefer deep, well-drained, moisture-retentive soil, but tolerate drought and winter wet. This species is among the most adaptable and easily grown of all crabapples; however, the scab resistance experienced at UBCBG may not be universal, nor its freedom from rust, mildew, and common apple pests.

PROPAGATION: Seeds require a period of cold stratification to overcome embryo dormancy. Seed-grown plants will show the species's inherent variation, or its propensity for hybridization if other *Malus* species are nearby. Cutting propagation is possible, but little information regarding this species is available. Most commercial *Malus* varieties are grafted or budded onto clonal rootstocks, and superior selections of *M. yunnanensis* could be propagated this way.

Meliosma pinnata var. *oldhamii* (Maximowicz) von Beusekom
Oldham worm-head tree
syn. *Meliosma oldhamii* Maximowicz
Sabiaceae   Plate 183

DISTRIBUTION: Southeast Asia.

A MOSTLY tropical genus of trees and shrubs, *Meliosma* displays either pinnately compound or simple leaves that are either evergreen or deciduous. There is a small number of moderately hardy, gardenworthy, but mostly untried species, and these are all deciduous. The genus name derives from Greek *meli*, meaning "honey," and *osma*, meaning "smell." While *Meliosma* flowers are indeed lovely, the most striking feature of the species is their leaves, which are handsome and pest free. The unusual common name probably derives from the twisted (worm-like) radicle that is characteristic of the genus.

Meliosma pinnata var. *oldhamii* is the most frequently cultivated member of the genus, but it is by no means commonly cultivated. It has odd-pinnate leaves (that is, with terminal leaflets), and is noted for its coarse, stiff branching and large panicles of tiny, fragrant flowers. The foliage resembles a small-leaf, alternately arranged *Fraxinus* (ash) or *Phellodendron* (cork tree), with green to blue-green shiny leaves more or less clustered on stout, silvery grey-green stems and broad, serrate leaflets often both folded and drooping. The rounded, compressed winter buds are scaleless but covered with rufous hairs, presenting an attractive contrast with the leaves and, in winter, with the stems. Plants are upright in youth, but given space, they will eventually form a spreading crown. Miniature cup-and-saucer flowers are produced in large terminal panicles once the leaves are nearly expanded; hanging reddish drupes follow. At close quarters one can detect a faint sour milk aroma emanating from the leaves, but it is not pervasive and is easily excused given the tree's many virtues.

The variety *oldhamii* was named for Richard Oldham (1837–1864), botanical collector for the Royal Botanic Gardens, Kew, who collected it in 1863 in the Korean islands and reported in 1867. His collection was then overshadowed by the Chinese collections of E. H. Wilson some 15 years later. The same plant is found in the southern Japanese islands and Taiwan.

This taxon differs little from *Meliosma pinnata*, except perhaps in winter hardiness, as the typical form of the species hails from decidedly tropical climes throughout Indo-Malaysia. Oldham worm-head tree grows slowly in the open at UBCBG, and has reached only 1.5 m in 10 years in this site. Another plant of the same age, grown in a shaded site at VanDusen Gardens some 10 km away, is more than 3 m tall. The tree is known to reach 20 to 40 m.

HARDINESS: Zone 8 or 7. Despite reports of limited tolerance for frost in Britain, *Meliosma pinnata* var. *oldhamii* appears hardy in the Pacific Northwest, probably because of the absence of late spring frost in this area.

CULTIVATION: This species is adaptable and trouble free in the garden, but prefers deep, moist soil and full sun to perform optimally.

PROPAGATION: Propagate from seed or layers. Viable seed is sometimes produced from isolated trees.

Meliosma veitchiorum W. B. Hemsley
Veitch worm-head tree
Sabiaceae   Plates 184–185

DISTRIBUTION: Western China.

LIKE *Meliosma pinnata* var. *oldhamii*, the western Chinese *M. veitchiorum* has odd-pinnate leaves (that is, with terminal leaflets) and is noted for its coarse, stiff branching and large panicles of tiny, fragrant flowers. It is a giant. Nearly every part is outsized and exceptionally bold, from the massive overwintering buds to the leaves and flower clusters, pendulous fruit stalks, and the cob marble-like fruits. Annual growth on rapidly growing shoots can reach 3 to 4 m, while the distance between nodes on reproductive shoots is approximately 1.5 m. Even young plants show the handsome grey-on-white vertically ridged and furrowed bark. Leaves can grow to nearly 1 m long and comprise 9 to 11 broad leaflets, each on a red stem and between 9 and 20 cm long. The bottommost leaflets are comparatively small but increase in size farther along the rachis, and the terminal leaflet is larger still. The winter outline of *M. veitchiorum* is admittedly spare and gaunt, but the felted buds, stout branches, and gigantic leaf scars are compelling features.

One of the UBC collections is derived from seed collected at Huang Shan, in Anhui Province, at 1377 m elevation. The parent tree itself is described as a giant multistemmed specimen 15 m across. Its exact location is on the Shi Ba Lo Han (18-Curve Pathway), close to Bei Hai (North Sea), so named because the view from the peak is usually of a sea of cloud filling the valley below.

HARDINESS: Zone 8 or 7. *Meliosma veitchiorum* is later to start into growth than *M. pinnata* var. *oldhamii* and is affected neither by late frosts nor deeper winter cold in the Pacific Northwest.

CULTIVATION: *Meliosma veitchiorum* is adaptable and trouble free in the garden, but prefers deep, moist soil and full sun to perform optimally.

PROPAGATION: Propagate from seed or layers. Viable seed is sometimes produced from isolated trees.

Melliodendron xylocarpum Handel-Mazzetti
Chinese parasol storax
syns. *Melliodendron jifungense* Hu, *M. wangianum* Hu
Styracaceae   Plates 186–187

DISTRIBUTION: Southern China.

THIS EXCITING introduction from China has only been cultivated in the West for a few years. It is a rare species, endemic to low-elevation valleys and wet forests at between 600 and 1500 m in China's southern provinces. Despite its warm-temperate provenance, plants at UBCBG are thriving, suggesting that the species may once have had a wider distribution and could possess considerable hardiness. *Melliodendron* is a monotypic genus closely related to *Sinojackia* and *Rehderodendron*, and is characterized by woody, pear- or spindle-shaped fruits; the epithet *xylocarpum* means "woody fruit."

Limited information on these trees in their native habitat exists, but they are described as growing 6 to 20 m tall. Judging from the growth rate and juvenile form of plants in cultivation, it appears that *Melliodendron xylocarpum* will become a small, delicately branched tree. In our garden, bark and shoots are dark grey-brown with sparsely set, bright green leaves that are ovate with a long drip tip, about 15 cm long, and are very late to drop in autumn. New growth, including twigs, flower stalks, and developing fruits, is initially covered with an attractive golden green, close-set pubescence.

This tree's most compelling feature is its widely opening (to 6 or 7 cm across) styrax-like flowers that emerge from fuzzy, conical buds. When fully open, they resemble little floating parasols among the young, pea-green foliage. The flowers are pink, fading to almost white, with a deep pink line down the centre of each broad petal. Flowering and leafing out in late March exposes these tissues to the vagaries of late winter weather. This time of year is generally frost free in the garden, but 2001 proved otherwise, and a number of early opening flowers, such as those of *Magnolia sprengeri*, were wrecked. Though they did not flower that year, the melliodendrons were undamaged.

HARDINESS: Possibly Zone 8, but the true hardiness of this species is unknown at this time. It is susceptible to late spring frosts.

CULTIVATION: Otherwise relatively pest and disease free in the Pacific Northwest, plants are intolerant of poorly oxygenated soils and best suited to woodland locations. Well-drained, deep, rich soil with summer moisture suits *Melliodendron*, as it does the majority of other styracaceous species.

PROPAGATION: Seedling production appears to be relatively easy, given the numbers of plants currently available from China. Cuttings are probably a viable alternative to seed, as the majority of related plants are readily propagated in this way.

Nyssa sinensis Oliver
Chinese tupelo
Nyssaceae

Plates 188–189

DISTRIBUTION: Central China

A HANDSOME, undemanding tree, *Nyssa sinensis* has distinctive, long, dark green leaves. Most seedlings adopt an excurrent crown habit (like a Christmas tree) with broad horizontal branches, and quickly become arborescent, unlike the behaviour of its American relatives, which, in our garden at least, often remain shrubby for many years. Like its American kin, though, Chinese tupelo is spectacular in its autumn repose and is worth planting for that feature alone. In most years, occasional leaves in the lower canopy often begin turning bright orange in midsummer, but following the autumn equinox, the mass of already dark leaves begins to take on a darker reddish hue and eventually turns to scarlet and purple as the green pigments are finally replaced.

Once mature, Chinese tupelo becomes broad headed, and tiny, female, greenish white flowers are produced in the leaf axils of younger branches in April. The flowers are followed in September by clusters of 2 or 3 oblong (6 mm by 12 mm) blue fruits. In nature, *Nyssa sinensis* grows in wet, mixed forest conditions along streams and valley bottoms in central China, where it grows to 20 m. Given their rapid growth rate in our garden, we expect the trees to reach at least that height in cultivation.

HARDINESS: Zones 7–9. As the species is distributed over a wide geographical and topographical range, provenance is probably a significant determinant in cold hardiness.

CULTIVATION: The species is somewhat shade tolerant, but in full sun it is more compact and it colours more spectacularly. The Chinese tupelos at UBCBG benefit from high overhead shade and moist soil; however, the species is used in street plantings in a number of Greater Vancouver municipalities and appears to be adjusting well to poorer, drier soils. In our garden, trees are remarkably pest and disease free, except for a single occurrence of damage by leaf-cutter bees (*Megachile* spp.). These insects are known to sample, then persevere annoyingly on a single species; one year they tried *Nyssa sinensis*, but they have not returned.

PROPAGATION: Ripe seed is easy to germinate and semiripe-wood stem cuttings root readily.

Photinia beauverdiana C. K. Schneider **Dashahe form**
syn. *Pourthiaea beauverdiana* (C. K. Schneider) Hatusima
Rosaceae   Plates 190–191

DISTRIBUTION: China.

IN A GENUS generally known for a single ubiquitous evergreen, *Photinia* ×*fraseri*, the deciduous photinias stand out in stark contrast. *Photinia beauverdiana* is one of a small number of deciduous species grown for their polished leaves that colour beautifully in autumn, and rounded corymbs of tiny white flowers in spring. *Photinia* is believed to have been named *photos*, the Greek word for "light," an allusion to the highly reflective leaves. The shining, small, orange to red, crabapple-like pomes, however, are arguably the most appealing feature of these plants. Naturally slim and upright in habit, perhaps to no more than 10 m, *P. beauverdiana* bears slender branches that eventually arch away and droop at the edges of the crown under the weight of these fruits.

The species and its variety *notabilis*, which bears slightly larger leaves and fruit, are both reasonably well known in gardens, particularly in Britain. Plants at UBCBG are quite different from those. They derive from a collection made by Peter Wharton in the Dashahe (dash-a-huh) Cathaya Reserve in Guizhou, China, from trees with extraordinarily large (to 15 mm), striking, vermilion to scarlet fruit. Otherwise, this variant closely resembles the typical *Photinia beauverdiana*, exhibiting narrow, pointed, papery leaves that turn spectacular orange before falling. The fruits themselves are abundantly produced and are quite tasty, which could reasonably indicate human involvement in the development of this particular form.

Botanists, especially in Asia, sometimes separate the deciduous photinias into the genus *Pourthiaea*. That name derives from the French missionary Pourthie. The species epithet *beauverdiana* commemorates Gustave Beauverd (1867–1940), a Swiss botanist and botanical artist.

HARDINESS: Zone 6, if the UBCBG selection is similar to the species.

CULTIVATION: Our plants grow quickly in well-drained, humus-rich soil and appear to tolerate summer drought; however, the region is generally cool and humid. Plants in the genus are known to prefer open situations or light shade, and the deciduous species are susceptible to fireblight.

PROPAGATION: Our plants were grown from seed and are remarkably uniform, suggesting the original plants may be apomictic (embryos formed directly from maternal tissue, hence seedlings are essentially clones). A short period of cold stratification is usually required for germination of the seeds.

Picea smithiana (Wallich) Boissier
Morinda spruce, Himalayan spruce
Pinaceae   Plate 192

DISTRIBUTION: The western Himalayas in Afghanistan, Nepal, northern India, and western Xizang (Tibet), to 3600 m.

THE MILD CLIMATE in the Pacific Northwest allows a wide range of plants to be cultivated, but also favours a number of ornamental plant pests. Two particularly serious pests are spruce aphid and spruce mite. Both of these arthropods are phloem feeders, active in the late winter and early spring on a number of *Picea* species, including the widely cultivated *P. pungens* (Colorado spruce) and *P. abies* (Norway spruce). In cold winter areas, low temperatures reduce overwintering populations of these pests, but where it is mild, they proliferate and eventually defoliate interior needles so that specimens look sick and scrubby. A number of *Picea* species are resistant and Morinda spruce is among these.

Few *Picea* species are as stately or striking as *P. smithiana* (Morinda spruce), one of only three spruces known for impressively pendulous branchlets; the other two are the slighter *P. breweriana* (Brewer spruce) from northern California and southern Oregon, and the naturally occurring pendulous forms of *P. abies*, sometimes referred to as var. *viminalis* (Alströmer) Liljeblad. Morinda spruce is a massive species, growing to 50 m or more in nature, but it is probably better known in gardens for its considerable spread than for its height. In cultivation, it easily produces a dense, 8-m-wide curtain of branches in as little as 25 years. Morinda spruce's sharp, pointed needles are long for a spruce (to 4 or 5 cm), and they spread radially around the shoot, pointing downward with a slight curve inward. The female cones are straight, smooth, and cylindrical, and the largest of the genus at 12–18 cm by 5 cm. Gradually turning from green to nut brown, they attractively stud the crown from summer onward.

HARDINESS: Zone 5. Although cold hardy, Morinda spruce starts into growth early in the season, so new growth may be damaged by late spring frosts in areas where they occur.

CULTIVATION: Other than its considerable lateral space requirement, *Picea smithiana* appears to be trouble free and to grow rapidly in cultivation, even in limestone soils.

PROPAGATION: Seed, which is copiously produced in the huge cones.

Picrasma quassioides W. E. Bennett
Picrasma, Bitter ash
syn. *Picrasma ailanthoides* Planchon
Simaroubaceae   Plates 193–194

DISTRIBUTION: Eastern Asia and the Himalayas.

A SMALL, often multistemmed deciduous tree that ranges over much of East Asia and the Himalayas, *Picrasma quassioides* boasts a number of desirable attributes. The species is known primarily in botanical collections, which is regrettable, as it is entirely suitable for gardens, growing to no more than perhaps 10 m tall in 25 years. The genus *Picrasma* includes both trees and shrubs, their alternately arranged leaves pinnately compound, and their bark and leaves characteristically bitter, as they contain quassin, a pharmacologically important compound found throughout the family. According to *King's American Dispensary* (1898), quassin imparts tonic properties in water or alcohol, making it valuable against various ailments including dispepsia, fevers, and drunkenness (although for drunkenness, "frequent bathing of the head in cold water is a valuable auxiliary"). Quassin is also a common food additive, imparting bitterness to a variety of beverages, food, alcoholic drinks, and lozenges. The name *Picrasma* comes from the Greek *picrazein*, meaning "to cause a bitter taste," and *quassioides*, "like Quassia," a similar plant from Surinam, named by Linnaeus for Graman Quasi, a slave who used the bark of that tree to treat fever.

Picrasma quasssioides displays handsome leaves, red-stalked in full sun, with 9 to 15 broad, serrated leaflets that change from deep green to yellow or orange, or, on drier sites, reddish in autumn. Leaves are clustered near the ends of the branchlets, much like those of the closely related *Ailanthus altissima* (tree of heaven), but the branching is finer and more elegant. And unlike that species, bitter ash has no unpleasant fragrance from bruised tissue or new growth. The bark is smooth, at least in youth, and dotted attractively with yellowish lenticels. Equally at home at the edge of woodland or in the open, *P. quassioides* appears to be an ideal candidate for urban situations. *Picrasma* produces flowers in the summer that are small and not particularly showy. In warm summer climates, vivid orange-pink berries appear following the flowers, nestled among the leaves in autumn, and gradually turn black as the leaves senesce.

HARDINESS: Zone 5, despite published reports that plants are only hardy in Zone 10. Like many other plants that have wide geographic ranges, provenance is probably important in the hardiness of individual plants.

CULTIVATION: Adaptable and fast growing in the Pacific Northwest, picrasma is reported to perform best on cooler, well-drained sites in neutral or acid soil.

PROPAGATION: Seed, cold stratified in autumn to overcome dormancy, germinates after 4 or 5 months.

Platycarya strobilacea Siebold & Zuccarini
Cone walnut
syn. *Fortunaea chinensis* Lindley
Juglandaceae   Plates 195–197

DISTRIBUTION: China, Japan, Korea, and Taiwan.

This unusual, handsome, deciduous tree is surprisingly rare in cultivation, despite being rather common in its native haunts. *Platycarya* is a monotypic genus, its single species displaying typical walnut family traits of separate-sexed flowers and pinnately compound leaves. It makes a shrub or tree to 12 m tall with a rounded and open-branched habit, nearly always attractive. Leaves are relatively small for a walnut, usually less than 30 cm long, composed of 7 to 15 ovate, slender-pointed, serrated leaflets. These are initially softly hairy, but later glabrous, as are the shoots, and alternately arranged toward the end of stout, spare branches.

Closely related to *Pterocarya* (wingnut), *Platycarya* differs in having solitary, upright, cone-shaped female inflorescences (the wingnuts are grouped in pendulous strings), each surrounded by a sizable cluster of upright male catkins. The ripened infructescence, composed of alternating, stiff, lanceolate bracts and winged nutlets, stands about 4 cm. It remains green until the leaves turn yellow and fall, when it gradually becomes biscuit coloured then dark brown, usually breaking apart before leaves emerge in the spring. The small, winged nutlets give the genus its name: from the Latin *platy*, meaning "broad," and *karyon*, meaning "a nut." The epithet *strobilacea* means "cone-like."

Hardiness: Zone 6, but given the species's substantial native range, there are probably more and less cold-hardy selections of the species.

Cultivation: Full sun and fertile, well-drained soil suits *Platycarya* at UBCBG. The species is known to have done poorly under similar conditions in the United Kingdom, suggesting problems with late spring frost.

Propagation: Viable seed is sometimes produced even on solitary trees and is readily germinated after a period of cold stratification. Plants can also be successfully layered in situ, and scions can be grafted onto *Carya* (hickory) understock.

Quercus acuta Thunberg
Japanese evergreen oak
syn. *Cyclobalanopsis acuta* (Thunberg) Oersted
Fagaceae   Plate 199

Distribution: Japan and Korea.

Oaks are universally known as massive, long-lived trees with tough, durable wood. Evergreen oaks are generally smaller versions, exhibiting similar strength and character, but their foliage, which is generally unlike that of the more common deciduous oaks, imparts a certain elegance and charm to a genus usually described in more powerful terms (for example, the mighty oak). For this and other reasons, evergreen oaks are a valuable commodity. It may seem odd that they are not commonly seen in the landscape, but the southern North American evergreen oaks (usually referred to as live oaks) tend to languish without considerable summer heat and do not generally tolerate cold winter temperatures.

A number of evergreen oaks native to Asia, however, show tremendous promise in both the cool Pacific Northwest and warmer areas of North America. This group, sometimes referred to under the name *Cyclobalanopsis*, comprises some 150 species, but few of them are hardy in the West and most have not been cultivated extensively here; so we do not actually know much about their hardiness.

The rarely cultivated Japanese *Quercus acuta* has the bearing of a southern magnolia. This species boasts beautifully polished, broadly ovate leaves without teeth, and an upright, spreading habit. In spring, the leaves and shoots are coated with a rusty indumentum, though this quickly wears away as the tissues expand. The small (2-cm) acorns are not produced on plants at UBC. Initially compact, unless shaded, *Q. acuta* eventually becomes arborescent, but like other evergreen oaks, it is most attractive when its lower branches are left intact, allowing it to maintain its shrubby character and only slowly form a low-branched tree. The dark grey-brown bark is smooth but dotted attractively with raised yellowish lenticels. The occasional transverse wrinkles below larger branches and on the trunk near the ground suggest (in miniature) the folded skin on elephant's feet. The epithet *acuta* is from the Latin for "sharp" and refers to the drawn-out (attenuate) leaf tips.

HARDINESS: Zone 8. Although not as hardy as *Quercus myrsinifolia*, this species also does well in the cooler Pacific Northwest and is likewise known to perform admirably in the southeastern United States.

CULTIVATION: Heat and drought tolerant, this oak requires full sun and good drainage and benefits from deep, rich soil. Shade-grown plants are drawn-up and will eventually decline unless allowed to emerge into full sun.

PROPAGATION: Seed, which requires no pretreatment when fresh, and semiripe-wood cuttings, are both effective techniques for raising plants.

Quercus myrsinifolia Blume
Bamboo leaf oak
syn. *Cyclobalanopsis myrsinifolia* (Blume) Oersted
Fagaceae   Plates 200–201

DISTRIBUTION: Japan, China, and Laos

PERHAPS the most elegant of the hardy Asian evergreen oaks yet tried in Western gardens is *Quercus myrsiniflolia*. Bamboo leaf oak is a small tree with a broad, rounded crown that can be exceptionally dense in youth. Without training, at about 20 years the crown begins to open up to show its smooth, silver-grey bark and graceful, well-spaced scaffold branches. Like *Q. acuta*, *Q. myrsinifolia* bears occasional transverse wrinkles below larger branches and on the trunk near the ground, suggesting (in miniature) the folded skin on elephant's feet. The shiny, barely serrated leaves are leathery, narrow, and pointed, and hang vertically in gentle arcs. Emerging bronze-red to purple brown, once they fully develop they turn mid green above and glaucous beneath, their posture and shape strongly suggestive of bamboo leaves or the leaves of myrtle. The epithet *myrsinifolia* derives from the Latin *myrsine*, the classical name for myrtle, and *folia*, meaning "leaf."

At UBCBG, trees produce pollen-bearing catkins and the wind-pollinated flowers typical of oaks, but acorns are never fully developed, suggesting that temperatures may be inadequate for that. The actual cause may be the absence of another pollen source for cross-pollination, however, as the species is commonly propagated by stem cuttings, not seed. Where acorns are formed, they are relatively narrow, about 2.5 cm long, with a hemispherical cup (cupule) marked with tiny concentric scales. Bamboo leaf oak is a common constituent of the evergreen broad-leaf forest in much of Southeast Asia, where it grows 12–25 m tall.

HARDINESS: Zone 7. Although this species is hardier than *Quercus acuta*, provenance of the original stock probably has much to do with adaptability and hardiness. It does well in the cooler Pacific Northwest and is also known to perform admirably in the southeastern United States.

CULTIVATION: Heat and drought tolerant, this oak requires full sun and good drainage and benefits from deep, rich soil. Shade-grown plants are drawn-up and will eventually decline unless allowed to emerge into full sun.

PROPAGATION: Seed, which requires no pretreatment when fresh, and semiripe-wood cuttings, are both effective techniques for raising plants.

Rehderodendron macrocarpum Hu
Rehder sausage tree
Styracaceae   Plates 201–202

DISTRIBUTION: Western China and Vietnam.

A SMALL GENUS of 9 species of spreading deciduous trees and shrubs, *Rehderodendron* is related to *Halesia* (silverbells). Like *Halesia* species, *Rehderodendron* species are known for their magnificent spring displays of fragrant, bell- or snowdrop-shaped flowers. The hardiest species (presumably) in cultivation, *R. macrocarpum* is native to the mountains of Sichuan, Yunnan, and adjacent Vietnam. Because the genus is primarily restricted to warm-temperate southern and western China, the species are not well known in the West; *R. macrocarpum* was only described by botanists in 1931 and introduced to cultivation in 1934. Like the silverbells, sausage tree flowers are pure white with prominent yellow stamens and are produced with the emerging foliage in May. The individual flowers of *Rehderodendron*, however, are even more deliciously fragrant, bigger, and borne in showier axillary panicles of up to 10 flowers. Like the majority of styracaceous trees, flowers are borne pendulously along the underside of horizontal branches, but the ascending branch tips often show the flowers to greater advantage in *Rehderodendron*.

Eventually forming a massive, sprawling crown with sinuous branching in the Pacific Northwest, *Rehderodendron* is a slow grower and often somewhat shrubby in youth. Few trees in cultivation are known to be larger than 10 or 12 m tall; a 40-year-old specimen at UBCBG is only 6 m tall by about 8 m wide. The large, dark green leaves are elliptic to lance shaped, cupped slightly, and have a characteristic drooping but forward-pointing posture. In the Pacific Northwest, these trees seem to be completely pest and disease free and the leaves are remarkably unblemished. Both the petioles and new shoots are a striking red, as are the strange, ribbed, sausage-shaped fruits that hang attractively along the branch undersides. In winter, the fruits litter the ground under mature specimens, where they slowly decompose, gradually releasing their charge of seeds.

The name *Rehderodendron* commemorates Alfred Rehder (1863–1949), one of America's most renowned botanists, the curator of the Herbarium of the Arnold Arboretum at Harvard University and an expert on Chinese plants and the world's temperate flora. *Dendron* means "tree," and the epithet *macrocarpum* means "large fruit."

HARDINESS: Zone 8.

CULTIVATION: Like other members of Styracaceae, sausage tree requires moist, well-drained soil. In full sun exposure, the broad, spreading habit would probably limit its use in smaller residential gardens, but shaded trees, while more shy to flower, are often strongly upright and narrow. In southern Yunnan, felling of old growth in the 1950s has resulted in regrowth from stumps, and vigorous multistemmed trees to 12 m high are common. Seed production among these trees is impressive and it appears that squirrels or other rodents collect and hoard the fruits as they do at UBCBG.

PROPAGATION: Propagation is relatively easy with seeds, but these are seldom produced without benefit of cross-pollination. Parthenocarpic fruit (without viable seeds) are produced from isolated specimens. While semiripe-wood stem cuttings are not difficult to root, the emergence of a good leading shoot often takes considerable time.

Sorbus commixta Hedlund **Mount Kurohime form**
Japanese mountain ash
Rosaceae   Plates 203–205

DISTRIBUTION: Korea, Japan, and Sakhalin.

FEW TREES are as beautiful yet as confusing to gardeners as *Sorbus commixta*, the Japanese mountain ash. The species is highly variable across its native range. Plants in cultivation under this name are mostly narrowly upright trees with coarse but attractive foliage. They are generally renowned for their small, orange-red berries and brilliant autumn colour; however, plants referable to this species include smaller, spreading, even shrubby trees. Appropriately, the specific epithet *commixta* means "mixed together." Luckily, all manifestations of the species appear to have ornamental value, and one, a small, multistemmed form growing at UBCBG, has caught the attention of a number of garden visitors. Originally from seed collected on the slopes of Mount Kurohime, a popular ski hill in central Japan, this tree is quite unlike other cultivated forms. While it has the shiny, narrow, conical, nearly hairless buds characteristic of *S. commixta*, its leaves are small (no more than 25 cm by 10 cm) and the leaflets narrow, long-pointed, and glaucous beneath.

The mature form of this Japanese mountain ash is diminutive, but definitely arborescent; single-stemmed individuals produce low arching branches, and multistemmed plants produce stiffer, more ascending branches. Plants grow to only about 4 or 5 m tall after 25 years. As a small garden tree, there is much to recommend it. Not only are its fine-textured leaves elegant, but its white flowers are produced in striking, plate-sized corymbs along the exposed upperside of the branches. Although individually small, the glistening, polished berries that follow are either pumpkin orange or startling vermilion red, and abundantly produced. Birds, normally drawn to mountain ashes like squirrels to oaks, leave these berries till December at least. Autumn leaf colour is also striking (orange and crimson), occurring often some weeks later than in other forms of the species.

HARDINESS: Zone 5, although some forms may exhibit varying degrees of cold hardiness due to differences in provenance.

CULTIVATION: Mountain ashes are notoriously short-lived in many regions. Borers can be troublesome in some areas and fireblight is often a limiting factor in areas where warm, moist conditions are coincident with plants flowering and leafing out. In the Pacific Northwest, cool spring conditions and wet winters reduce or even eliminate these problems completely; nevertheless, some fireblight is still seen on highly susceptible species. Luckily, *Sorbus commixta* is not among these. Here, Japanese mountain ash is easily grown in sun or semishade, in almost any soil.

PROPAGATION: Seed-grown plants from cultivated sources will almost inevitably be hybrids if other mountain ash species are present (that is, within pollinator flying distance); there is almost no place left that European mountain ash (*Sorbus aucuparia*) is not a common tree, for example. Seedlings from single, isolated trees or from multiple plants of the same species will closely resemble the parent plant. Selected individuals are often budded onto seedling *S. aucuparia* rootstocks.

Sorbus hupehensis C. K. Schneider
Hubei mountain ash
syn. *Sorbus glabrescens* (Cardot) Handel-Mazzetti
Rosaceae   Plates 206–207

DISTRIBUTION: China.

THE MOUNTAIN ash is one of the most recognizable trees in urban North American plantings, and most keen gardeners know that there are both pinnately compound-leaf species (such the two species described in this book), as well as simple-leaf species (often known as whitebeams). *Sorbus* species are among the earliest of deciduous plants to leaf out, and fresh, unfolding leaves accompany the flowers, adding an attractive contrasting element to flowering.

Despite there being a good number of native species across the continent, most people know best the exotic *Sorbus aucuparia* (European mountain ash). The rowan, as it is known in the United Kingdom, is a moderate-sized, hardy, short-lived tree noted for its masses of attractive orange to red berries. For the average North American gardener, familiarity with this species is not adequate preparation for either of the two Himalayan species outlined here.

As with the majority of mountain ashes, *Sorbus hupehensis* produces small white flowers in flattened corymbs at the shoot tips. Although it is similar to *S. oligodonta*, *S. hupehensis* is easily distinguished once its flowers fade and fruit begins to ripen. Hubei mountain ash produces small, stark white berries in loose clusters; these stalks are an arresting rose pink and hang on even after the berries have disappeared in mid-winter. The species is strong-growing, but ultimately forms a medium-sized, compact tree with glaucous, blue-green, pinnately compound leaves on polished purple-brown stems. The leaflets are oblong and finely serrated for most of their length, which distinguishes them from *S. oligodonta*. Fall colour is usually in shades of orange and red.

The pinnate-leaf Himalayan species are notoriously difficult to sort out, as many are extremely similar and they reproduce by apomixis (that is, they produce viable seed without fertilization). The seedlings of apomictic species are therefore identical—they are clones of the parent plant. The form of *Sorbus hupehensis* we see in gardens appears to reproduce exclusively by apomixis. In cultivation, such trees often vary from place to place, but differences are probably explained by the environmental conditions peculiar to each site. Low humidity and high light levels appear to produce more intensely coloured fruit, bluer leaves, and narrower, more compact growth than would be produced under cooler, cloudy conditions, for example. Not surprisingly, these tendencies have encouraged a proliferation of names, particularly for "selected" seedlings of the pink-fruited species. For example, the cultivars 'Coral Fire', 'November Pink', 'Pink Pagoda', 'Rosea', and 'Rufus' probably represent the same clone.

HARDINESS: Zone 5.

CULTIVATION: An excellent garden plant, *Sorbus hupehensis* prefers deep, well-drained, moist soil for best growth, but also tolerates poorer, less open soil. Maximum sun exposure to the crown is ideal, as plants tend to grow loosely, produce little fruit, and have green leaves when shaded. The bark, however, is relatively thin and easily damaged by intense winter sunlight or by mechanical injury.

PROPAGATION: Berries are generally produced in abundance on mature trees, although birds may make short work of them if they are still on the tree much beyond Christmas. The seeds require a short period of cold stratification for best germination.

Sorbus oligodonta (Cardot) Handel-Mazzetti
Kite-leaf rowan
syn. *Sorbus hupehensis* var. *obtusa* C. K. Schneider
Rosaceae   Plates 208–209

DISTRIBUTION: Southwestern China and Myanmar (Burma).

LIKE MOST mountain ashes, *Sorbus oligodonta* produces small white flowers in flattened corymbs at the shoot tips. It bears heavy masses of larger white berries that appear dusted with crimson rouge. The red intensifies as the weather turns cold, but eventually fades as the berry expands and the skin stretches out, so that by about Christmas, the berries are pearly white. These berries distinguish *S. oligodonta* from *S. hupehensis*. Kite-leaf rowan is strong-growing, but ultimately makes a medium-sized, compact tree. It has glaucous, blue-green, pinnately compound leaves on polished purple-brown stems. The leaflets are more obovate, or kite shaped, than those of *S. hupehensis*, with only a small number of teeth near the apex. The leaves colour well in autumn, usually in shades of orange and red. The species name *oligodonta* translates as "few teeth." *Sorbus oligodonta* appears to reproduce exclusively by apomixis. (For further details, see discussion under *S. hupehensis*.)

HARDINESS: Zone 5.

CULTIVATION: This excellent garden plant prefers deep, well-drained, moist soil for best growth, but also tolerates poorer, less open soil. Maximum sun exposure to the crown is ideal, as plants tend to grow loosely, produce little fruit, and have green leaves when shaded. The bark is relatively thin and easily damaged by intense winter sunlight or by mechanical injury.

PROPAGATION: Berries are generally produced in abundance on mature trees, although birds may make short work of them if they are still on the tree much beyond Christmas. The seeds require a short period of cold stratification for best germination.

Sorbus yuana Spongberg
Yuan whitebeam
syn. *Aria yuana* (Spongberg) H. Ohashi & I. Oketani   Plates 210–212

DISTRIBUTION: Western China.

AT FIRST GLANCE, many people are misled by the simple-leaf *Sorbus* species, the whitebeams, thinking them *Crataegus* (hawthorn), *Malus* (crabapple), or *Pyrus* (pear), to which they are closely allied. Vegetatively, they do not much resemble their pinnate-leaf kin (the mountain ashes), but ignoring this, there are obvious affinities in buds, flowers, and fruits. In truth, this group of genera, part of the Maloideae, a subfamily of the rose family, is problematic, with botanists seldom agreeing on the best way to interpret the species. Some botanists separate the whitebeams into their own genus, *Aria* (Persoon) Host, and make a good case for doing so, but this approach has not yet caught on generally. Two species of white-beam are somewhat familiar to gardeners: *S. aria* (European whitebeam) with its initially grey-green, softly hairy leaves and large, spotted, red berries, and *S. alni-folia* (Korean whitebeam), a variable species with alder-like leaves and slighter bunches of mostly smaller, spherical, pink to red fruit. *Sorbus yuana* is most similar to the latter species.

Rare in the wild and one of the rarest whitebeams in cultivation, *Sorbus yuana* is a small to medium-sized spreading tree to 15 m, with tomentose young shoots and leaves. The finely serrated leaves are elliptic (approximately 10 cm by 6 cm) and prominently veined with 11 to 13 nearly parallel lateral veins, impressed in the glossy, dark green, upper leaf surface. Below, the leaves are greyish white and later, pale green. Emerging early, they are seemingly the perfect foil for the pale flowers. In autumn the leaves change to a sumptuous flame orange, even in semi-shade. In May in our garden, magnificent corymbs up to 20 cm across are produced on elongated shoots. These presage the arrival of huge clusters of 1.5-cm-diameter lozenge-shaped berries. The fruits are, technically, pomes—all species in the apple subfamily produce such fruits—and although initially pale green, they ripen in October and November to a startling vermilion. Yuan whitebeam is native to western Hubei and Sichuan, along steep slopes above 2000 m elevation. The late Theodore "Ted" Dudley, a U.S. National Arboretum research taxonomist who collected our material, once mentioned that the local people snack on the berries pulled from the lower limbs as they hike through the forest.

HARDINESS: The cold hardiness of *Sorbus yuana* is unknown, but given its native range and the known hardiness of its close relatives, it is probably hardy to Zone 6.

CULTIVATION: Yuan whitebeam is a strong grower in the Pacific Northwest. Trees are easily transplanted and undemanding as long as they are provided with good drainage. Growth rate and ultimate size are related to shade and the availability of moisture. A full-sun exposure will provide maximum flowering and fruiting.

PROPAGATION: Seed propagation is relatively easy, if a period of cold stratification is given to break dormancy. Most trees will produce viable seed from self-pollinations; but *Sorbus* species are notoriously promiscuous and may produce hybrid seed if other species, either mountain ash or whitebeam, are in close proximity.

Styrax hemsleyana Diels
Hemsley snowbell
Styracaceae   Plate 213

DISTRIBUTION: Central China.

INTRODUCED to the West by Ernest H. Wilson in 1900, *Styrax hemsleyana* is virtually unknown outside of botanical collections, which is not only curious but a serious oversight. There are few outstanding small trees suitable for residential gardens, and fewer still that have the refined bearing of Hemsley snowbell. In cultivation, the species attains a height of no more than 6 or 7 m and is slow growing, which may explain why commercial nurseries are often less than excited about such introductions. Each of its horizontal branches is well spaced and tapers gently upward to a slender tip. The dark grey-brown bark is notably smooth, and the buds are chocolate brown. Flowers are typical for *Styrax*—fragrant, splayed white bells about 2.5 cm across, with prominent yellow anthers—and produced on short pedicels in long racemes or panicles up to 15 cm long. Appearing in June, the blossoms are a beautiful complement to the fully expanded, heavily textured, light green leaves. The leaves, flower stalks, and shoots are initially covered in a copper-coloured pubescence that gives the tree a subtle green softness. Later in the summer, fruits form: grey-yellow, flannel-covered, 1.5-cm-diameter globes on fine, wire-like stalks. They hang in bunches behind the leading shoots like minia-ture Christmas tree decorations.

In some respects Hemsley snowbell is intermediate between the more famil-iar *Styrax japonicus* and *S. obassia*. Its broadly elliptic, pubescent leaves and flow-ers borne in lax, terminal inflorescences suggest *S. obassia*, but it is not as prone to shrubbiness nor as vigorously upright and the leaves are smaller and less coarsely toothed. The habit of the tree is compact and similar to, but more open-branched and, I would argue, more graceful than *S. japonicus*. Native to mountain slopes and forest edges below 1000 m elevation in central China, it is known to grow 5–12 m tall. The name *Styrax* is the classical Greek name for these plants; the epithet com-memorates William Botting Hemsley (1843–1924), keeper of the Kew Herbarium.

A number of other Chinese *Styrax* species have come to light since the late 1990s, and many of these probably warrant closer examination. One species in particular, *S. tonkinense*, looks particularly distinctive and beautiful. In our gar-den, it began flowering profusely at an early age with slender, fragrant white bells emerging from elongated, cinnamon-brown buds. Time will tell whether this spe-cies is hardy and worth recommending.

HARDINESS: Zone 6, although *Styrax hemsleyana* is often rated as being more ten-der. Seedlings require extra protection, as young roots are considerably less cold hardy than their aboveground parts.

CULTIVATION: *Styrax hemsleyana* requires an acid, well-drained soil to which organic matter has been added. Like other *Styrax* species, it will grow in the open in sun, but usually performs better with a cool root run and the kind of protec-tion afforded by high overhead shade. Early summer moisture is important, espe-cially while plants are growing and flowering, but once established, these plants are surprisingly resilient (except where there is compaction of the root zone). *Styrax* species require some coddling in youth and are generally unhappy if kept in containers for long periods.

PROPAGATION: A mature tree produces copious amounts of seed, and most of it will germinate if sown fresh following 3–5 months of warm stratification, followed by 3 months of cold stratification. Plants can also be propagated from stem cut-tings and probably by grafting onto *S. japonicus* seedlings.

Taiwania cryptomerioides Hayata
Coffin tree
syn. *Taiwania flousiana* Gaussen
Cupressaceae   Plates 214–215

DISTRIBUTION: Western China, Myanmar (Burma), northern Vietnam, and Taiwan.

Coffin tree is a rare conifer that displays a number of attractive and interesting features, not least that the exceptionally light wood was traditionally used in making coffin boards. A monotypic genus, in its native range specimens of *Taiwania* have been recorded at over 70 m tall. Although few examples of any great size now exist in the wild, coffin tree is known to be an emergent species in mixed forests in monsoon-affected areas. Chinese plants are sometimes referred to the species *T. flousiana*, but differences between continental and Taiwanese populations are evidently not botanically significant. In cultivation, Chinese trees are reported to have longer, greener, softer leaves. *Taiwania* is neither well known nor widely cultivated in the West, primarily because the species was thought to lack winter hardiness, coming as it does from the subtropics; but more than 20 years of successful cultivation at UBCBG and other gardens in the Pacific Northwest and the United Kingdom proves otherwise.

The prickly juvenile leaves of *Taiwania* are awl-shaped and forward pointing on the shoot, resembling the foliage of the related and more garden-familiar *Cryptomeria* (the renowned *sugi* of Japan). After some years, *Taiwania* begins producing specialized terminal branches with smaller, more closely appressed, scale-like leaves. Cones are borne on these shoots. Female cones are small and oblong with smooth, rounded, overlapping cone scales, something like those of *Tsuga* (hemlock), though the coffin tree's closest relatives are thought to be *Cunninghamia* (Chinese fir) and *Athrotaxis* (Tasmanian cedar). Curiously, seedlings of *Taiwania* and *Cunninghamia* are nearly indistinguishable, while their mature foliage and cones are distinct; and coning shoots of *Taiwania* closely resemble *Athrotaxis* or *Sequoiadendron* (Sierra redwood), but their juvenile forms are completely unalike.

Coffin trees in the forested David C. Lam Asian Garden show the typical conical habit, beautiful blue foliage, and graceful pendulous branch tips of the species as it grows in the wild. In the sun, plants are invariably bushy and much slower growing than those given overhead protection, and a number of our specimens are dwarfed and produce multiple leaders. Some are chronically sunburned in the open, but a few are doing well, despite the exposure. None of the trees in our garden yet exhibits cones or the distinctive dimorphic shoots that bear them.

Hardiness: Zone 8 or perhaps hardier. Late-spring frost and lack of overhead protection are probably significant obstacles to cultivation. Chinese forms of this species are reported to be the hardiest. The UBCBG collection is comprised entirely of high-elevation (2200 m) Taiwanese accessions.

Cultivation: Trees are reported to grow slowly in cultivation and in the wild, but *Taiwania* appears to revel in the forested conditions at UBCBG, growing to 7 m or more in 20 years. Although our thin, acid soils were amended with organic matter before planting and supplemental irrigation is available, specimens do not seem bothered by extensive summer drought, in spite of the species's monsoon provenance. Our oldest plants still retain a skirt of branches to the ground, even in shade.

PROPAGATION: *Taiwania* cuttings are easily rooted with time, but the terminals of side shoots are slow to develop apical dominance, so plants with multiple leaders should be utilized if they are available. To date, seed has not been available from cultivated sources.

Tetradium glabrifolium (Champion ex Bentham) Hartley
Hong Kong evodia
syn. *Euodia glabrifolia* Champion ex Bentham
Rutaceae   Plates 216–217

DISTRIBUTION: Southeast Asia.

ONCE KNOWN under the genus name *Evodia*, a name later discarded for the more accepted spelling *Euodia* (the name is Latin for "sweet or pleasant scent"), the temperate Asian pinnate-leaf species were split off into the genus *Tetradium*. Perhaps as a result of such nomenclatural upheavals, these plants are known better for their confused taxonomy than their beauty as garden plants. Both *T. glabrifolium* and *T. hupehensis* are considered by some to be merely variants of the widespread Asian *T. daniellii*, for example, and while the two are distinct in our garden, descriptions for these taxa vary widely and do not always match with cultivated material. *Tetradium* is closely related to *Phellodendron*, but the species are not pungently aromatic as they are in the latter genus.

    Tetradium glabrifolium is a smaller, twiggier tree than *T. hupehensis* and it requires considerable heat, coming as it does from a more southern range than *T. hupehensis*. Hong Kong evodia has not set fruit in our garden, as it is usually still in flower at Halloween and does not completely drop its leaves until December. Consequently, fall colour is poor, but the late-flowering, large, persistent, red-violet inflorescence stalks more than make up for this. The species tends to be rounded in habit—nearly every branch on our tree has a pronounced curvature—and forms a low-branched, bushy tree approximately as wide as it is tall (5 m tall and wide in 15 years).

HARDINESS: To Zone 8, at least.

CULTIVATION: Hong Kong evodia is easily cultivated in sun or light shade in any but poorly drained soil. Pest and disease problems are few at UBCBG; problems affecting rutaceous plants are more common in warmer summer climates. The species is heat tolerant.

PROPAGATION: Propagation can be achieved by seed (no pretreatment is required) and semiripe-wood cuttings. Root cuttings are also reported to be successful.

Tetradium hupehensis (Dode) T. G. Hartley

Hubei evodia, Chinese bee tree

syns. *Evodia hupehensis* Dode, *Tetradium daniellii* (Bennett) T. G. Hartley
　　Hupehense Group

Rutaceae   Plate 218

DISTRIBUTION: China.

A SPECIMEN of *Tetradium hupehensis* at UBCBG has grown more than 15 m tall in 20 years. Planted in shaded woodland, the tree has a crown that is admittedly narrower than it would be in the open, but its well-spaced branches are clean and give it a handsome winter outline. The leaves are oppositely arranged and shortly pinnately compound, each with 5 to 9 lustrous leaflets. The bark is smooth and grey. In late summer, small white flowers are abundantly produced in broadly rounded terminal panicles. Each flower produces copious nectar, and trees in bloom are mobbed by bees and other pollinating insects. Tiny, orange, ultimately reddish capsular fruits follow, and upon ripening in autumn, the capsules split open to reveal pairs of shiny black seeds. When viewed at close range, the seeds can be seen to vibrate in the air, as they are attached by only a short, hair-like funiculus. At that time, leaves begin to yellow, creating an attractive contrast to the reddish brown infructescences.

HARDINESS: Zone 6 or 5.

CULTIVATION: Chinese bee tree is easily cultivated in sun or light shade in any but poorly drained soil. The species has few pest and disease problems at UBCBG and is heat tolerant.

PROPAGATION: Propagation can be achieved by seed (no pretreatment is required) and semiripe-wood cuttings. Root cuttings are also reported to be successful.

Noted Collectors of Asian Plants

Cox, Euan Hillhouse Methven (1893–1977). The son of a Scottish jute merchant, Euan Cox developed a notable garden at Glendoick in his later years. His collecting started when noted plantsman Reginald Farrer invited him on an expedition in Upper Burma in 1919. He returned later that year, and Farrer remained to continue collecting in the field, but died there shortly after. Cox then assumed the unhappy duty of sorting out their joint collections, and the experience is written up in his book, *Farrer's Last Journey* (1926). Cox also wrote the classic, *Plant Collecting in China* (1945). Euan's son Peter and grandson Kenneth have continued the family tradition of plant collecting in China.

David, Jean Pierre Armand (1826–1900). Born in Espelette, France, Père David served as a missionary in China from 1861, where he documented numerous plants, a deer (Père David deer), and the giant panda. Eponymy includes *Davidia involucrata*, *Acer davidii*, and *Buddleja davidii*.

Delavay, Pierre Jean Marie (1834–1895). Born in Haute-Savoie, France, Père Delavay was a Jesuit missionary who explored the region around Canton and southwest Yunnan. Eponymy includes *Abies delavayi*, *Aster delavayi*, *Osmanthus delavayi*, and *Thalictrum delavayi*.

Farges, Paul Guillaume (1844–1912). Père Farges was born in Monclar-de-Quercy in southern France. He served as a missionary in China from 1867 until his death. Between 1892 and 1903, he collected extensively in northeast Sichuan and sent seed to French nurseryman Pierre Vilmorin. Eponymy includes *Abies fargesii*, *Decaisnea fargesii*, *Paulownia fargesii*, and *Rhododendron fargesii*.

Farrer, Reginald (1880–1920). Farrer was a Yorkshireman who, after graduating from Oxford University, settled briefly in Japan. From his base in Tokyo, he explored China, Japan, and Korea. After returning to England, he mounted an expedition to Tibet and Gansu (China), described in his book *On the Eaves of the World* (1917). He died at Nyitadi while plant collecting on his final, fateful expedition to Upper Burma (see also E. H. M. Cox, above). Eponymy includes *Gentiana farreri* and *Viburnum farreri*.

Forrest, George (1873–1932). A Scot from Falkirk, Forrest trained as an apothecary, then went to work at the Royal Botanic Garden in Edinburgh. His first trip to China was funded by A. K. Bulley, the founder of Bees Seeds in England. In all, Forrest visited China on five gruelling expeditions. On the last of these he became ill and died in the upper Mekong valley of Yunnan. Bulley's garden with its Forrest material later became the basis for the Ness Botanic Gardens of Liverpool University, United Kingdom. Eponymy includes *Pieris forrestii*, and *Rhododendron forrestii*.

Fortune, Robert (1812–1880). Born in Edrom, Berwickshire, United Kingdom, Fortune trained at the Royal Botanic Garden in Edinburgh and then took a position as hothouse superintendent for the Royal Horticultural Society in London. He set off for China in 1843 to collect for the society, and worked around Shanghai and in the Wuyi Shan. On a junk travelling from Foochow to Chusan, he was attacked by pirates and only escaped with his life after defending the junk with considerable heroism. He returned to London in 1846, but in 1848 set off for China again, making four trips in all. His fifth trip (1860–1862) was to Japan, a country that was just opening after the treaty of Kamegawa had been signed in 1854. Eponymy includes *Euonymus fortunei* and *Mahonia fortunei*.

Handel-Mazzetti, Heinrich (1882–1940). A notable Austrian collector during the First World War, Handel-Mazzetti was active in that part of western China where the Salween, the Yangtze, and the Mekong Rivers carve deep gorges out of the mountains. In 1914 he went to China under the auspices of the Vienna Academy of Sciences and, due to the outbreak of hostilities, was not able to return to Europe until 1919. He used his enforced exile well, collecting widely over a huge area, and his travels are chronicled in his book, *Naturbilder aus Südwest China* (1927). His botanical specimens are preserved in the Vienna Natural History Museum. Eponymy includes *Arisaema handelii*.

Henry, Augustine (1857–1930). Born in Portglenone, County Antrim, Ireland, and trained in medicine, Henry went to work for the Imperial Custom Service in China. From this official position, he was able to collect living material for the Royal Botanic Gardens, Kew, and he published an account of the flora of Taiwan. His most famous work was his seven-volume *Trees of Great Britain and Ireland*, a collaboration with Henry John Elwes. He later became Professor of Forestry at the Dublin College of Science. Eponymy includes *Emmenopterys henryi*, *Lilium henryi*, *Saruma henryi*, and *Viburnum henryi*.

Hooker, Sir Joseph Dalton (1817–1911). Hooker was one of the greatest botanists of all time. The son of Sir William Hooker, at age 22 he sailed with Captain Ross on the HMS *Erebus* and HMS *Terror* voyage to the southern oceans. He returned from the frozen south full of insights into plant geography, which later made him sympathetic to the ideas of his friend Charles Darwin. He determined next to gain experience in tropical botany, and in 1847 he set out to explore the Sikkim Himalaya for the Royal Botanic Gardens, Kew. From his base in Calcutta, he spent two years of adventurous exploring, which he wrote up in *Himalayan Journals* (1854). He later (like his father) became director of the Royal Botanic Gardens, Kew. Eponymy includes *Berberis hookeri* and *Rhododendron hookeri*.

Kingdon Ward, Francis (Frank) (1885–1958). The son of Cambridge botanist Harry Marshall Ward, as a boy Frank overheard one of his father's colleagues say, "There are places up the Brahmaputra where no white man has ever been," a chance comment that was to determine his career. From 1910 to 1956, he searched for new and beautiful plants in the mountains of Tibet, China, Myanmar, and India. Caught in a terrible earthquake in the Lohit

Gorge, he saw the mountain forests "peeled off like wet paper." His introductions include *Meconopsis betonicifolia* and *Primula florindae* (named for his wife), and more than twenty rhododendrons. Eponymy includes *Acer wardii* and *Meconopsis wardii*.

Lancaster, Roy (1937–). Horticulturist, traveller, writer, and broadcaster, Lancaster has collected plants extensively in East Nepal, China, and Russia. A native of Bolton in the north of England, he gained early experience in tropical botany during two years serving with the British army in Malaya (1956–1958). After training at the University of Cambridge Botanic Garden, he began working for Hillier Nurseries in 1962, becoming in 1970 the first curator of the Hillier Arboretum.

Maximowicz, Carl Johann (Karl Ivanovitch) (1827–1891). Maximowicz was a Russian botanist, born in Tula, who worked at the St. Petersburg Botanical Gardens. He travelled extensively in Mongolia and Japan. He set off on an expedition to East Asia in 1854 and brought plants back to the St. Petersburg Botanical Gardens. He explored Japan extensively in the 1860s, aided by his collector Chonosuke Sugawa. He corresponded with Charles Sargent at the Arnold Arboretum, and through this association many plants from Japan reached North America via St. Petersburg. Eponymy includes *Acer maximowiczii*, *Alnus maximowiczii*, *Betula maximowicziana*, *Picea maximowiczii*, and *Populus maximowiczii*.

Meyer, Frank N. (1875–1918). Born Frans Meijer in Amsterdam, Meyer worked as a boy at the Amsterdam Botanical Gardens, where he caught the attention of the great Hugo de Vries who encouraged him further in botany. He emigrated to America, arriving in 1901, and began working for the U.S. Department of Agriculture (USDA). At the request of David Fairchild and the USDA Foreign Plant Introduction Section, he made his first visit to China from 1905 to 1908. His final visit to China commenced in 1916 with the aim of collecting *Pyrus ussuriensis* and *P. calleryana*, both reputedly resistant to fireblight disease. In Jingmen he was able to collect 2300 kilograms of pear fruits for seed. During this time, strain and loneliness were taking their toll, as evidenced in his letters to

Fairchild. While travelling by steamer down the Yangtze, on his return from Hankow to Shanghai, he tragically drowned near Wuhu. It is not known whether he fell off the ship accidentally or committed suicide in a fit of depression. Eponymy includes *Syringa meyeri*.

Ogisu, Mikinori (fl. 1983). A Japanese horticulturalist, scholar, and traveller, Ogisu has botanized extensively in China, on occasion with Roy Lancaster. He has been particularly active in tracking down *Rosa chinensis* var. *spontanea*, a supposed major breeding source of the modern rose, in its wild state in China. In 1983 he found it at Leibo in southwest Sichuan. He later found it in both northern Sichuan and in ravines of the Emei Shan. Eponymy includes *Epimedium ogisui*.

Perny, Paul-Hubert (1818–1907). Abbé Perny was initially a parish priest in Besançon, France, before becoming a missionary in Sichuan. A general naturalist, he had time while in Sichuan to document (in addition to many plants) Perny's long-nosed squirrel (*Dremomys pernyi*) and the Chinese oak silkmoth (*Antheraea pernyi*). Eponymy includes *Disporopsis pernyi* and *Ilex pernyi*.

Rock, Joseph Francis Charles (1884–1962). An able botanist, anthropologist, and adventurer, Rock was one of the most daring explorers of the Tibetan borderlands of China. Born in Vienna, he taught himself Chinese characters at age thirteen. From 1907 to 1920, he lived in Hawaii, becoming an expert on the flora there. In the 1920s he went to China, where he lived until 1949. A great linguist, he played Caruso to astonished Naxi villagers on a wind-up gramophone, while studying their language and culture. Rock sent thousands of pressed plants and seeds to his American backers, such as Charles Sargent at the Arnold Arboretum and the U.S. Department of Agriculture. Eponymy includes *Brighamia rockii* and *Paeonia rockii*.

Sargent, Charles Sprague (1841–1927). Son of a wealthy Boston merchant, Sargent was educated at Harvard College and served in the Union Army. After touring gardens in Europe, he became active in horticultural circles in New England. In 1872 Harvard's botanic garden, the Arnold Arboretum, was established, with Sargent as founding director.

In his position there, he worked tirelessly to raise money for funding expeditions to China, Korea, Japan, and Tibet, including, among others, Ernest H. Wilson and Joseph Rock. The plants brought back by his collectors established the preeminence of the Arnold Arboretum for its Asian collections. Eponymy includes *Malus sargentii*, *Prunus sargentii*, and *Viburnum sargentii*.

Siebold, Phillip Franz von (1796–1866). Siebold was a German doctor, born in Wurzburg, who in search of adventure signed up with the Dutch East India Company, initially to be stationed in Batavia (Jakarta). The Dutch soon sent him to Japan, where he arrived in 1823 at Dejima, a Dutch trading concession in this otherwise closed nation. He quickly began collecting what plants he could get hold of, which were among the first ever to reach Europe from Japan. Among other ventures Siebold opened a European-style school for Japanese students in Nagasaki. Eponymy includes *Primula sieboldii*.

Sugawa, Chonosuke (1841–1925). Sugawa was assistant to Carl Maximowicz, in the botanizing of Japan in the 1860s. The Russianized orthographic variant of his name is Tschonoski Sukawa, which explains the Latin form of plants named after him. Eponymy includes *Acer tschonoskii*, *Lonicera tschonoskii*, *Malus tschonoskii*, and *Rhododendron tschonoskii*.

Wilson, Ernest Henry (1876–1930). Born in the Cotswolds of England, Wilson worked for Hewitt Nurseries at Solihull and subsequently at the Birmingham Botanical Gardens, the Royal Botanic Gardens Kew, and the Royal College of Science in South Kensington, London. At the invitation of Veitch and Sons, the famous nursery, he set off to explore China, as he was later to do for Charles Sargent at the Arnold Arboretum. His activities in China earned him the sobriquet "Chinese Wilson." He also visited Japan to collect cherry trees for the Arnold Arboretum. In 1927 he became curator of that arboretum. Three years later, he died in a car accident in Worcester, Massachusetts. He wrote the classic book on Chinese gardening, *China, Mother of Gardens* (1929). Eponymy includes *Corydalis wilsonii* and *Ilex wilsonii*.

Conversion Tables

To convert:	Multiply by:
Millimetres to inches	0.04
Centimetres to inches	0.4
Metres to feet	3.3
Kilometres to miles	0.62
Square kilometres to square miles	0.38
Kilograms to pounds	2.2
Hectares to acres	2.5

Temperatures

$$°C = 5/9 \times (°F{-}32) \qquad °F = (9/5 \times °C) + 32$$

Average Annual Minimum Temperature for Each Zone

°Celsius		Zone	°Fahrenheit	
−46	and below	1	−50	and below
−43	to −46	2a	−45	to −50
−40	to −43	2b	−40	to −45
−37	to −40	3a	−35	to −40
−34	to −37	3b	−30	to −35
−32	to −34	4a	−25	to −30
−29	to −32	4b	−20	to −25
−26	to −29	5a	−15	to −20
−23	to −26	5b	−10	to −15
−20	to −23	6a	−5	to −10
−18	to −20	6b	0	to −5
−15	to −18	7a	5	to 0
−12	to −15	7b	10	to 5
−9	to −12	8a	15	to 10
−7	to −9	8b	20	to 15
−4	to −7	9a	25	to 20
−1	to −4	9b	30	to 25
2	to −1	10a	35	to 30
4	to 2	10b	40	to 35
4	and above	11	40	and above

Glossary

acuminate drawn out into a long, thin point.

adaxial the side toward the axis (stem) (e.g., the upper surface of a leaf).

apex, apices tip of an organ or shoot.

apomictic reproducing asexually, either by vegetative parts (e.g., runners or bulbils) or by asexually produced seed (agamospermy).

aril a fleshy edible seed-coat, often red or orange, functioning to attract seed-dispersing animals.

armature general term for the protective outgrowths on the surface of a plant (spines, thorns).

attenuate narrowed.

bullate surface puckered with small, rounded convexities.

campanulate bell-shaped.

cauliform stem-like.

cline continuous geographically based variation in morphology.

coriaceous leathery.

cordate heart-shaped.

corymb, corymbose flat-topped, racemose inflorescence (i.e., flowers tending to the same level. See also *umbel*.

crenate (of leaves) with blunt, rounded teeth.

cuneate wedge-shaped.

cyme, cymose important general type of inflorescence in which the inflorescence terminates in a single flower, but branches below that flower with branches also terminating in a single flower, and so on. Compare *raceme*.

dehiscent opening spontaneously when mature or ripe (as a fruit or anther).

drupe fleshy, single-seeded fruit with a stone (e.g., plum).

epiphyllous situated on leaves.

epiphyte plant growing on another.

Eocene a geological epoch 54 million years ago.

exfoliation peeling off in strips or flakes.

follicle dry, many-seeded seedpod derived from a single carpel, splitting along one edge.

fourmerous four-parted. See also *tetramerous*

geosynclinal pertaining to broad, elongated depressions in the earth's crust containing great thicknesses of sediment.

hirsute hairy.

hyperhumid with very high rainfall and humidity.

indole butyric acid (IBA) a synthetic auxin-like plant growth hormone, commonly used to promote the growth of cuttings.

indumentum general term for the hair covering of a plant.

inflorescence cluster of flowers.

karst eroded limestone landscape.

lenticels gas-exchange pores of bark usually protected by corky cells and often a prominent feature of young bark.

mesophytic of or pertaining to plant growth in non-extreme conditions, i.e., with adequate water and heat.

Miocene geological epoch from 23.8 to 5.3 million years ago.

monotypic (of a genus) comprised of a single species

obovate egg-shaped but with the widest part at the tip. Compare *ovate*.

orographic pertaining to physical relief, e.g., mountains).

ovate egg-shaped with the widest part at the base. Compare *obovate*.

panicle loose cluster of flowers formed by the branching of a raceme.

pectinate comb-like.

pedicel stalk of an individual flower. Compare *peduncle*.

peduncle the stalk holding an inflorescence.

perlite expanded volcanic rock often used for propagation composts.

phylogeny the relationships of groups of organisms as reflected by their evolutionary history.

physiographic pertaining to geographic features of the earth's surface.

Pleistocene recent geological epoch characterised by glacial periods from 1.8 to 0 million years ago, including the present, although in some definitions, the past 10,000 years is distinguished as the Holocene.

Pliocene geological epoch from 5.3 to 1.8 million years ago.

polymorphic taking many forms.

raceme, racemose inflorescence with flowers borne on a common central stalk, thus usually elongated; not terminating in a single flower. Compare *cyme*.

rachis midrib.

radicle the embryonic root, its emergence through the seedcoat usually the first outward sign of germination in a seed.

reticulate net-like.

rhombic diamond-shaped.

rugulose surface with small bumps.

scandent climbing or clambering.

sclerophyllous vegetation characterized by scrubby trees and shrubs with leaves that are evergreen, small, hard, thick, and leathery.

ternate (of leaves) compound and divided into three or more parts.

Tertiary period of geological time from 65 to 1.8 million years ago, including the Miocene and Pliocene epochs.

tetramerous (of flowers) with parts (stamens, petals, etc.) in fours.

tomentose covered with matted, short, soft hairs.

umbel, umbellate flat-topped cymose inflorescence (i.e., flowers tending to the same level). Compare *corymb*.

verticillaster a ring of flowers produced at a node, in fact formed of two opposite axillary cymes, as in many members of the mint family.

xeric of or pertaining to plant growth in dry conditions such as deserts. Compare *mesophytic*.

Bibliography

An Z., W. Pinxian, W. Shumig, et al. 1991 Changes in the monsoon and associated environmental changes in China since the last interglacial. In T. Liu, ed. *Loess, Environment, and Global Change, Series of the 8th INQUA Congress*. Beijing: Science Press.

Axelrod, D. I., I. Al-Shehbaz, and P. H. Raven. 1996. History of the modern flora of China. In A. Zhang and S. Wu, eds. *Floristic Characteristics and Diversity of East Asian Plants, Proceedings of the IFCD*. Beijing: China Higher Education Press; Berlin: Springer Verlag. 43–55.

Azegami, Chikara, ed. 1996. *Wildflowers of Japan: Mountainside*, Mountain Handy Picture Guide 2. Tokyo, Japan: Yama-kei Publishers.

Bai, P. Y., ed. 2000. *Flora Yunnanica: Tomus 11, Spermatophyta*. Beijing: Science Press.

Bailey, L. H. 1976. *Hortus Third*. Rev. by staff of the L. H. Bailey Hortorium, Cornell University. New York: MacMillan.

Bartholomew, B., D. E. Boufford, and S. A. Spongberg. 1983. *Metasequoia glyptostroboides*: its present status in central China. *Journal of the Arnold Arboretum* 64: 105–128.

Bean, W. J. 1976. *Trees and Shrubs Hardy in the British Isles*, 8th ed. London: John Murray.

Bean, W. J. 1988. *Trees and Shrubs Hardy in the British Isles*, Supplement by D. L. Clarke. London: John Murray.

Boufford, D. E., and S. A. Spongberg. 1983. Eastern Asian–eastern North American phytogeographical relationships: A history from the time of Linnaeus to the twentieth century. *Annals of the Missouri Botanical Garden* 70: 423–439.

Bretschneider, E. 1898. *History of European Botanical Discoveries in China*, 1962 reprint of the original 2-vol. ed. Leipzig: Zentral Antequariat der Deutschen Demokratischen Republik.

Bright, C. 1998. *Life out of Bounds: Bioinvasions in a Borderless World*, Worldwatch Environmental Alert Series. New York: W. W. Northon

Chamberlain, D. F., G. C. G. Argent, et al. 1988. *The Rhododendron Handbook 1998, Rhododendron Species in Cultivation*. London: The Royal Horticultural Society.

Chang, H. 1981. The Quinghai–Xizang Plateau in relation to the vegetation of China. In *Geological and Ecological Studies of Quinghai–Xizang Plateau, Proceedings of the Symposium on Quinghai–Xizang Plateau*. Beijing: Science Press. 1897–1903.

Cox, P. A. 1990. *The Larger Rhododendron Species*. London: Batsford.

Cribb, P. 2003. A botanist in Guizhou. *The Plantsman* n.s. 2: 142–151.

Cronk, Q. C. B., and J. L. Fuller. 2001. *Plant Invaders: The Threat to Natural Ecosystems*. London: Earthscan.

Cullen, J. L. 1981. Microfossil evidence for changing salinity patterns in the Bay of Bengal over the last 20,000 years. *Paleogeography, Paleoclimatology, and Paleoecology* 35: 315–356.

Davis, C. C., P. W. Fritsch, J. Li, and M. J. Donoghue. 2002. Phylogeny and biogeography of *Cercis* (Fabaceae): Evidence from nuclear ribosomal ITS and chloroplast ndhF sequence data. *Systematic Botany* 27: 289–302.

Dirr, M. A. 1998. *Manual of Woody Landscape Plants*, 5th ed. Champaign, Illinois: Stipes.

Dirr, M. A. 2002. *Trees and Shrubs for Warm Climates*. Portland, Oregon: Timber Press.

Dirr, M. A., and C. W. Heuser Jr. 1987. *The Reference Manual of Woody Plant Propagation: From Seed to Tissue Culture*. Athens, Georgia: Varsity Press.

Farjon, A. 1998. *World Checklist and Bibliography of Conifers*. Kew, England: Royal Botanic Gardens.

Farrer, R. 1917. *On the Eaves of the World*. London: Arnold.

Fiala, J. L. 1988. *Lilacs: The Genus Syringa.* London: Batsford.

Flora of Taiwan Editorial Committee. 1998. *Flora of Taiwan*, vol. 4, 2d ed. Taipei: Department of Botany, National Taiwan University.

Fryer, J. 1996. Undervalued versatility. *The Garden* 121(II): 146, 709–714.

Global Witness. 2003. *A Conflict of Interests: The Uncertain Future of Burma's Forests.* London: Global Witness.

Grey-Wilson, C. 2000. *Clematis: The Genus.* Portland, Oregon: Timber Press.

Griffiths, M. 1994. *Royal Horticultural Society Index of Garden Plants.* Portland, Oregon: Timber Press.

Guan, K. Y., ed. 1998. *Highland Flowers of Yunnan.* Kunming, China: Yunnan Science and Technology Press.

Hansen, R., and F. Stahl. 1993. *Perennials and Their Garden Habitats*, 4th ed. Portland, Oregon: Timber Press.

Hayashi, Y., and C. Azekami. 1985. *Nihon no Jyumoku* (Woody Plants of Japan). Tokyo: Yama-kei Publishers.

He Shan-an, ed. 1998. *Rare and Precious Plants of China.* Shanghai, China: Shanghai Scientific and Technical Publishers.

Hillier Nurseries. 1998. *The Hillier Manual of Trees and Shrubs.* Newton Abbot, England: David and Charles.

Hinkley, D. J. 2003. A plantsman's observations on the genus *Hydrangea. Davidsonia* 14: 31–40, 48–58. Online at: www.ubcbotanicalgarden.org/davidsonia.

Hootman, S. 2003. Plant hunting along the Nu Jiang (Salween River) in Yunnan Province, China. In G. Argent and M. McFarlane, eds. *Rhododendrons in Horticulture and Science, International Rhododendron Conference 2002.* Edinburgh, Scotland: Royal Botanic Garden. 247–258.

Hunt, D. R. 1960. *Persea ichangensis. Curtis's Botanical Magazine* 173(I): 5–8.

Huxley, A., ed. 1992. *The New Royal Horticultural Society Dictionary of Gardening.* London: Macmillan.

Jacobsen, A. L. 1996. *North American Landscape Trees.* Berkeley, California: Ten Speed Press.

Jarvis, D. I. 1993. Pollen evidence of changing Holocene monsoon climate in Sichuan Province, China. *Quaternary Research* 39: 325–337.

Justice, C. L. 1984. Some observations on the flora of Emei Shan, with reference to rhododendrons at different altitudes. *Rhododendron Notes Records* 1: 117–135.

Justice, D. 2002. Snake bark maples at UBC Botanical Garden. *Davidsonia* 13: 44–48, 57–58. Online at: www.ubcbotanicalgarden.org/davidsonia.

Justice, D. 2003. Silver firs in UBC Botanical Garden. *Davidsonia* 14: 71–78. Online at: www.ubcbotanicalgarden.org/davidsonia.

Kingdon Ward, F. 1949. Plant hunting in Manipur (1). *Journal of the Royal Horticultural Society.* 74(7): 288–340.

Kingdon Ward, F. 1956. *Return to the Irrawaddy.* London: A. Melrose.

Kingdon Ward, F. 2001. *Frank Kingdon Ward's Riddle of the Tsangpo Gorges: Retracing the Epic Journey of 1924–25 in South-East Tibet.* K. Cox, ed. London: Antique Collectors Club.

Krüssman, G. 1985. *Manual of Cultivated Conifers.* Portland, Oregon: Timber Press.

Kubitzki, K., and W. Krutzsch. 1996. Origins of East and Southeast Asian plant diversity. In A. Zhang and S. Wu, eds. *Floristic Characteristics and Diversity of East Asian Plants, Proceedings of the IFCD.* Beijing: China Higher Education Press; Berlin: Springer Verlag. 56–70.

Lancaster, R. 1989. *Travels in China, A Plantsman's Paradise.* London: Antique Collectors Club.

Li, H., and H. J. Noltie. 1997. Miscellaneous notes on the genus *Paris. Edinburgh Journal of Botany* 54(3): 351–353.

Liew, P., C. Kuo, and M.-H. Tseng. 1995. Vegetation of northern Taiwan during the last glacial maximum as indicated by new pollen records. *Abstracts, 14th INQUA Congress.* Berlin: INQUA. 161.

Liew, P., C. Kuo, S. Y. Huang, and M.-H. Tseng. 1998. Vegetation change and terrestial carbon storage in eastern Asia during the Last Glacial Maximum as indicated by a new pollen record for central Taiwan. *Global and Planetary Change* 16–17: 85–94.

Liu, B., ed. 1996. *Atlas of China* (English ed.). Beijing: China Cartographic Publishing House.

Liu, K.-B. 1986. Pleistocene changes in vegetation and

climate in China. In *Abstracts, American Quaternary Association Conference*. 94.

Liu, K.-B. 1991. Quaternary vegetational history of the monsoonal regions of China. In *Special Proceedings INQUA XIII Congress, Beijing*. Berlin: INQUA. 111–113

Liu, T. 1985. *Loess in China*, 2d ed. Beijing: China Ocean Press.

Lord, T., ed. 2003. *Royal Horticulture Society Plant Finder 2003–2004*. London: Dorling Kindersley.

Mabberley, D. J. 1997. *The Plant Book*, 2d ed. Cambridge, England: Cambridge University Press.

Mabberley, D. J., D. E. Jarvis, and B. E. Juniper. 2001. The name of the apple. *Telopea* 9(2): 421–430.

Macdonald, A. B. 1986. *Practical Woody Plant Propagation for Nursery Growers*. Portland, Oregon: Timber Press.

Manos, P. S., and M. J. Donoghue. 2001. Progress in Northern Hemisphere phytogeography: An introduction. *International Journal of Plant Sciences* 162: S1–S2.

Meyer, F. G., and E. H. Walker, eds. 1965. *Flora of Japan*. Washington, DC: Smithsonian Institution.

Mithen, S. 2002. *After the Ice: A Global Human History, 20,000 to 5,000 BC*. London: Weidenfeld and Nicholson.

Myers, J. H., and D. R. Bazely. 2003. *Ecology and Control of Introduced Plants*. Cambridge, England: Cambridge University Press.

Ngamriabsakul, C., M. F. Newman, and Q. C. B. Cronk. 2000. Phylogeny and disjunction in *Roscoea* (Zingiberaceae). *Edinburgh Journal of Botany* 57(1): 39–61.

Noltie, H. 1995. New Irises from Yunnan. *New Plantsman* 2(3): 131–140.

Noltie, H. 2000. Dwarf yellow lilies from the Sino-Himalaya. *New Plantsman* 7(1): 19–20.

Noltie, H. J. 1994. *Flora of Bhutan*, vol. 3, Part 1: Royal Botanic Garden, Edinburgh.

Ohwi, J. 1965. Araceae. In F. G. Meyer and E. H. Walker, eds. *Flora of Japan*, trans. Tetsuo Koyama. Washington, DC: Smithsonian Institution. 255–264.

Polunin, O., and A. Stainton. 1984. *Flowers of the Himalaya*. New Delhi, India: Oxford University Press.

Poor, J. M., ed. 1984. *Plants that Merit Attention, Volume 1—Trees*. Portland, Oregon: Timber Press.

Rangsiruji, A., M. F. Newman, and Q. C. B. Cronk. 2000. A study of the infrageneric classification of *Alpinia* (Zingiberaceae) based on the ITS region of nuclear DNA and the trnL-F spacer of chloroplast DNA. In K. L. Wilson and D. A. Morrison, eds. *Monocots: Systematics and Evolution*. Collingwood, Australia: CSIRO. 695–709.

Rangsiruji, A., M. F. Newman, and Q. C. B. Cronk. 2000. Origin and relationships of *Alpinia galanga* (Zingiberaceae) based on molecular data. *Edinburgh Journal of Botany* 57(1): 9–37.

Rehder, A. 1940. *Manual of Cultivated Trees and Shrubs Hardy in North America*, 2d ed. New York: MacMillan.

Rushforth, K. 1987. *Conifers*. London: Christopher Helm.

Rushforth, K. 1999. *Trees of Britain and Europe*. London: HarperCollins.

Salisbury, H. E. 1989. *The Black Dragon Fire*. Boston: Little Brown.

Schopmeyer, C. S., ed. 1974. *Seeds of Woody Plants in the United States*. Agriculture Handbook 450, U.S. Forest Service, Washington, DC: U.S. Department of Agriculture.

Shupe, J. F., M. B. Hunsiker, J. F. Dorr, and O. Payne. 1995. *National Geographic Atlas of the World*, 6th ed. Washington, DC: National Geographic Society.

Spanner, T. W., H. J. Noltie, and M. Gibbons. 1997. A new species of *Trachycarpus* (Palmae) from west Bengal, India. *Edinburgh Journal of Botany* 54(2): 257–259.

Stearn, W. T. 1996. *Stearn's Dictionary of Plant Names for Gardeners*. London: Cassell.

Straley, G. B. 1986a. *Dipteronia sinensis*. *Pacific Horticulture* 48: 38–39.

Straley, G. B. 1986b. Field notes—*Trochodendron aralioides*. *American Nurseryman* 164: 114.

Straley, G. B. 1989. *Rehderodendron macrocarpum*. *Pacific Horticulture* 50: 42–44.

Straley, G. B. 1991. Presenting *Sinocalycanthus chinensis*—Chinese wax shrub. *Arnoldia* 51: 18–22.

Straley, G. B. 1992. *Trees of Vancouver*. Vancouver, British Columbia: UBC Press.

Straley, G. B. 1993a. *Alangium platanifolium*, a little known Asian shrub. *Washington Park Arboretum Bulletin* 55: 17–19.

Straley, G. B. 1993b. New and underutilized Asian trees for North American gardens. *Journal of Arboriculture* 19: 250–254.

Straley, G. B. 1994a. Asian plant collections at the University of British Columbia Botanical Garden. *Plant Collections Newsletter Canada* 5: 6.

Straley, G. B. 1994b. Bold-leaved Asian perennials. *Washington Park Arboretum Bulletin* 57: 14–17.

Straley, G. B. 1994c. Eastern treasures. *American Nurseryman* 180: 120–126.

Straley, G. B. 1994d. Field notes—*Lindera obtusiloba. American Nurseryman* 180: 110.

Straley, G. B. 1995. Notes on uncommon Asian trees. *Public Garden* 10: 36–37.

Straley, G. B. 1996. Himalayan jewelweed. *Gardens West* 10(2): 58–59.

Synge, P. 1980. *Lilies.* London: B. T. Batsford.

Valder, P. 1999. *The Garden Plants of China.* Portland, Oregon: Timber Press.

Van Gelderen, D. M., P. C de Jong, and H. J. Otterdoom. 1994. *Maples of the World.* Portland, Oregon: Timber Press.

Vidakovic, M. 1991. *Conifers: Morphology and Variation.* Trans. by Maja Soljan. Croatia: Graficki Zavod Hrvatske.

Wang, Chi-Wu. 1961. *The Forests of China,* Harvard University Publication Series No. 5. Petersham, Massachusetts: Maria Moors Cabot Foundation.

Wen, J. 2001. Evolution of eastern Asian–eastern North American biogeographic disjunctions: A few additional issues. *International Journal of Plant Sciences* 162: S117–S122.

Winkler, M. G., and P. H. Wang. 1993, Late Quaternary vegetation and climate of China. In H. E. Wright et al. *Global Climates since the Last Glacial Maximum.* Minneapolis: University of Minnesota Press.

Wu, Z., and P. Raven, eds. 2001. *Flora of China.* Beijing: Science Press; St. Louis, Missouri: Botanical Garden Press.

Wu, Z., and S. Wu. 1996. A proposal for a new floristic kingdom (realm). In A. Zhang and S. Wu, eds. *Floristic Characteristics and Diversity of East Asian Plants, Proceedings of the IFCD.* Beijing: China Higher Education Press; Berlin: Springer Verlag. 3–42.

Wyman, D. 1965, *Trees for American Gardens.* New York: MacMillan.

Yeo, P. F. 1992. *Hardy Geraniums.* Portland, Oregon: Timber Press.

Ying, T., Y. Zhang, and D. A. Buford. 1993. *The Endemic Genera of Seed Plants of China.* Beijing: Science Press.

Zhou, W., et al. 1992. Variability of monsoon climate in East Asia at the end of the last glaciation. *Quaternary Research* 39: 219–229.

Zonneveld, K. A. F., et al. 1997. Mechanisms forcing abrupt fluctuations of the Indian Ocean summer monsoon during the last deglaciation. *Quaternary Science Review* 16: 187–201.

Index